Glacier Bay Oceanographic Monitoring Program Analysis of Observations, 1993–2009

Natural Resource Technical Report NPS/SEAN/NRTR—2012/527

Seth L. Danielson

Institute of Marine Science
School of Fisheries and Ocean Science
University of Alaska Fairbanks
Rm. 114 O'Neill Building
Fairbanks, Alaska 99775-7220

January 2012

U.S. Department of the Interior
National Park Service
Natural Resource Stewardship and Science
Fort Collins, Colorado

The National Park Service, Natural Resource Stewardship and Science office in Fort Collins, Colorado publishes a range of reports that address natural resource topics of interest and applicability to a broad audience in the National Park Service and others in natural resource management, including scientists, conservation and environmental constituencies, and the public.

The Natural Resource Technical Report Series is used to disseminate results of scientific studies in the physical, biological, and social sciences for both the advancement of science and the achievement of the National Park Service mission. The series provides contributors with a forum for displaying comprehensive data that are often deleted from journals because of page limitations.

All manuscripts in the series receive the appropriate level of peer review to ensure that the information is scientifically credible, technically accurate, appropriately written for the intended audience, and designed and published in a professional manner.

Data in this report were collected and analyzed using methods based on established, peer-reviewed protocols and were analyzed and interpreted within the guidelines of the protocols. This report received informal peer review by subject-matter experts who were involved with its collection, but were not directly involved in the analysis or reporting of the data.

Views, statements, findings, conclusions, recommendations, and data in this report do not necessarily reflect views and policies of the National Park Service, U.S. Department of the Interior. Mention of trade names or commercial products does not constitute endorsement or recommendation for use by the U.S. Government.

This report is available from Southeast Alaska Network Inventory & Management website (http://science.nature.nps.gov/im/units/sean/oc_main.aspx) and the Natural Resource Publications Management website (http://www.nature.nps.gov/publications/nrpm/).

Please cite this publication as:

Danielson, S. L. 2012. Glacier Bay oceanographic monitoring program analysis of observations, 1993–2009. Natural Resource Technical Report NPS/SEAN/NRTR—2012/527. National Park Service, Fort Collins, Colorado.

NPS 132/112453, January 2012

Contents

	Page
Figures	v
Tables	vii
Appendices	ix
Abstract	xi
Acknowledgments	xii
Introduction and Background	1
Methods	3
Sampling and Field Protocol	3
Results	7
Regional Environmental Setting	7
Mean Monthly Conditions, Anomalies and Trends	17
Relation to Other Observed Data and Climate Indices	31
Discussion and Conclusions	33
Literature Cited	35

Figures

Page

Figure 1. Oceanographic sampling station locations are depicted with numbers and symbols. 3

Figure 2. Chart of sampling coverage from 1993 to 2009, depicting sample month and measurement type. 5

Figure 3. Mean atmospheric pressure at sea level contours for the North Pacific in January, April, July and October. 7

Figure 4. Idealized circulation map showing annual precipitation rates (vertical bars) and the large-scale oceanic circulation features in the Gulf of Alaska. 8

Figure 5. Monthly climatology of atmospheric conditions recorded at Juneau, Alaska over 1992–2009 (left column) and mean monthly T and S values averaged over 0–10 m (Ta, Sa) and 195–205 m (Tb, Sb) depth strata at Glacier Bay Station 04 (right column). 9

Figure 6. Annual cycle of the northeastern Gulf of Alaska Upwelling Index (left) and the coastal fresh water discharge (right), after Royer (1982). 10

Figure 7. Monthly anomalies of sea level pressure (SLP), temperature (T), wind speed (WS), and precipitation (P) recorded at Juneau (PAJN) over 1992–2009 11

Figure 8. Normalized monthly anomalies of the following major climate indices: the Pacific Decadal Oscillation (PDO) sea surface temperature index, the North Pacific Gyre Oscillation (NPGO) sea level height index, the El Niño-Southern Oscillation (ENSO) sea surface temperature index, and the Pacific-North American (PNA) sea level pressure index. 12

Figure 9. Structure of the PDO (left two panels) and ENSO (right two panels) modes of SST variability. 13

Figure 10. Spatial pattern of the NPGO (upper left) along with time series of salinity, nitrate and chlorophyll fluorescence fluctuations along Line P (lower) and in the CalCOFI observation area (right). 13

Figure 11. Maps for January, April, July and October of the PNA pattern (left) and its correlation with temperature (center) and precipitation (right). 14

Figure 12. Late winter/early spring ocean conditions (March to May) between the Gulf of Alaska and the northwest reaches of Glacier Bay. 16

Figure 13. Late summer/early fall ocean conditions (August to September) between the Gulf of Alaska and the northwest reaches of Glacier Bay. 16

Figures (continued)

Page

Figure 15. Station 04 cross-sections of the mean monthly temperature and salinity fields (first and third rows, respectively) along with companion depth-time anomaly plots (second and fourth rows, respectively). 18

Figure 16. Mean monthly time series of temperature (top), salinity (middle) and sigma-0 (bottom) for 45 m (red and blue), 100 m (black) and 250 m (gray) depth strata at Stations 01 and 04. 19

Figure 17. Time series of temperature and salinity anomalies at select depths. 21

Figure 18. Station 12 upper 50 m cross-sections of the mean monthly temperature and salinity fields (first and third rows respectively) along with companion depth-time monthly anomaly plots (second and fourth rows respectively). 21

Figure 19. Station 20 upper 50 m cross-sections of the mean monthly temperature and salinity fields (first and third rows respectively) along with companion depth-time monthly anomaly plots (second and fourth rows respectively). 23

Figure 20. Anomalies of temperature (T') vs. anomalies of salinity (S') at 0 m depth for all stations. 24

Figure 21. Anomalies of temperature (T') vs. anomalies of salinity (S') at 100 m depth for all stations. 25

Figure 22. Scatter plots of Station 20 temperature anomalies at 5 m depth against 12 atmospheric, oceanographic and climate time series. 32

Tables

                                                                                                                          Page

Table 1. Correlation table of temperature anomalies with depth at Stations 01, 04, 12 and 20 27

Table 2. Correlation table of salinity anomalies with depth at Stations 01, 04, 12 and 20 28

Table 3. Correlation of temperature at 10 m and 150 m depth levels between all station pairs 29

Table 4. Correlation of salinity at 10 m and 150 m depth levels between all station pairs 30

Appendices

	Page
Appendix A. Time-depth sections of temperature and salinity seasonal cycle and anomalies for the period of record over the upper 50 m.	37
Appendix B. Time-depth sections of temperature and salinity seasonal cycle and anomalies for the period of record over the whole water column.	61
Appendix C. Time series of temperature and salinity anomalies at select depths.	85
Appendix D. Temperature and salinity anomalies at select depths.	133
Appendix E. Regressions of temperature and salinity anomalies with atmospheric, oceanographic and climate time series.	139
Appendix F. Tables of temperature and salinity correlations at each station between select depth levels.	155
Appendix G. Tables of temperature and salinity correlations between stations at select depth levels.	163

Abstract

The Glacier Bay oceanographic monitoring data archive period of record (July 1993 to October 2009) is newly re-analyzed in this report. This report is intended to be a comprehensive reference manual for temperature and salinity at each monitoring station location; detailed plots for each station are provided in the Appendices. Mean annual cycles are characterized at all stations and across the water column. These climatologies are used to generate monthly anomalies, which are subsequently evaluated for the presence of long-term trends.

Temperature fluctuations appear to be regionally coherent whereas the salinity field exhibits more local variations from one portion of Glacier Bay to another. A few stations exhibit marginally statistically significant trends (at the 95% confidence level) in cooling or freshening over the period of record at some depth levels, but these trends are not representative of the bay waters as a whole. Analysis of the density structure on either side of Sitakaday Narrows suggests that near-bottom water renewal of the central basin occurs most readily between November and May, while intermediate water renewal likely occurs in summer months as well.

Comparison of the anomaly time series with large-scale climate indices (PDO, ENSO, NPGO and PNA) indicates that these atmospheric and oceanographic patterns of variability exhibit only weak connection with the variability observed within Glacier Bay. Correlations with variability measured at oceanographic station GAK1 (located in the northern coastal Gulf of Alaska) are similarly weak, suggesting that the location of Glacier Bay measurements are well suited to complement the GAK1 time series in providing an independent perspective on coastal Gulf of Alaska oceanic variability.

Acknowledgments

The privilege of newly analyzing a long time series of environmental conditions is not an opportunity that presents itself very often. I am indebted to those who have collected the data that make up the Glacier Bay oceanographic archive. Because of the length of the period of record that this dataset now spans, I have been able to undertake analyses that were not previously possible. I thank B. Moynahan, W. Johnson, and L. Sharman for their dedication to ensuring the highest possible quality in the data archive and to G. Eckert for my introduction to Glacier Bay.

Introduction and Background

Glacier Bay National Park and Preserve hosts a rare combination of pristine wilderness, continually evolving biological, glacial, and geological systems, a highly productive marine ecosystem, and the institutional infrastructure necessary to carry out and maintain ongoing scientific research programs. Oceanographic research cruises to Glacier Bay were first carried out between in the mid 1960's; the modern era of regular observations using high accuracy conductivity-temperature-depth (CTD) instrumentation began in 1993.

Despite the extensive research carried out within and near Glacier Bay over the course of the last century, there are relatively few process-oriented studies or publications that focus on the controlling dynamics of Glacier Bay's marine physics and marine ecosystem dynamics. Early work focused on basic characterization of the fjord's circulation (although without the benefit of direct current measurements) and seasonal property distributions (Pickard 1967, Matthews and Quinlan 1975). Matthews (1981) concluded that deepwater renewal takes place primarily between November and April but a more recent report by Hooge and Hooge (2002) suggests that deepwater renewal events may occur at any time of year. This issue remains pertinent today because understanding the underpinning of Glacier Bay's marine ecosystem dynamics requires a mechanistic description of the processes controlling the influx of nutrient-rich water and its subsequent residence time, utilization rates, and return pathways. Glacier Bay oceanographic monitoring, which began in 1993, led to an appreciation for inter-annual variability of the near-surface and sub-surface temperature and salinity fields. Etherington et al. (2004) found that seasonal anomalies of near-surface temperature and salinity are related to anomalies in day length, air temperature and precipitation. Hill et al. (2009) showed that the strongly seasonal discharge cycle (average range of 300–1000 m^3/s) is likely modulated by different discharge cycles from high elevation vs. low elevation watersheds. While Matthews and Quinlan (1975) focused on the role of glacial ice meltwater in maintaining the high levels of salinity stratification, Hill et al. (2009) show that peak discharge events might reach as high as 15,000 m^3/s and are primarily resultant from new precipitation and melting of the seasonal snow pack.

Oceanographic monitoring efforts in Glacier Bay were formalized with a sampling protocol that called for seasonal (quarterly) observations (Hooge et al. 2003). Following a 2006 program review (Etherington 2006), the National Park Service (NPS) Inventory and Monitoring Program (I&M) committed to continuing the oceanographic sampling and a subsequent 2009 protocol working group discussed program intent, implementation, and available resources (see Appendix E of Danielson et al. 2010). Following the template of Oakley et al. (2003), the original protocol was updated by Danielson et al. (2009), sent out for peer review and published as Danielson et al. (2010) and referred to hereafter as OC2010.1. OC2010.1 more closely aligns the oceanography sampling effort with the program objectives listed in Etherington (2006). The most substantive changes from Hooge et al. (2003) to OC2010.1 include the following: 1) a reduction of the spatial resolution of sampling except in mid-summer (July) and mid-winter (December/January), 2) an increase of the temporal resolution with monthly sampling of a select subset of stations between March and October, 3) the addition of one deep (150 m bottom depth) station outside of the fjord and the removal of two stations in Geike Inlet, and 4) the incorporation of a more fully automated and standardized system of data processing, error checking and archiving.

The sampling instrumentation employed in the monitoring effort is a Seabird, Inc. CTD data logger. Salinity (derived from the conductivity, pressure and temperature measurements), temperature, and pressure are used to compute water density; these parameters form the basis of the physical system monitoring carried out in this program. Ancillary sensors on the CTD (photosynthetically available radiation [PAR], fluorescence, optical backscatter [OBS], and dissolved oxygen [DO]) provide information about biologically important parameters. These parameters, however, are less easily assessed in the integrative or time series approaches employed in this document because of the various qualifications associated with each measurement. For example, PAR levels are impacted by the time of day, clouds, and sampling vessel shadows. Fluorescence, turbidity and oxygen measurements are not calibrated with *in-situ* analyses of seawater and so provide at best relative levels of each parameter. Different species of phytoplankton can have different responses to activation by fluorometers operating at different wavelengths so profiles at stations that host different phytoplankton species assemblages may provide different profiles that are not solely related to the concentration of fluorescing phytoplankton or chlorophyll-a concentration. The relative values are useful measures, but care must be taken to view them in proper context and with an understanding of their limitations. Comparisons among seasons and years are limiting for the same reasons. Furthermore, some of these parameters exhibit highly non-linear seasonal responses (e.g., spring bloom timing or coastal discharge suspended sediment load) that are highly susceptible to temporal aliasing. Thus, this report devotes relatively little attention to these secondary measurements.

This report is intended to be a reference manual for each monitoring station's physical properties (including "legacy" stations that are no longer part of the observation program). Data analyses and graphical depictions are presented to highlight both the similarities in properties and property fluctuations between stations as well as the differences that define each station's unique attributes. With well over a decade of sampling effort in many calendar months, a station-by-station temperature (T) and salinity (S) climatology is computed. Each station's monthly climatology enables a baseline property field from which standard anomalies can be computed and evaluated. We do not attempt to incorporate historical Glacier Bay measurements (mostly from the 1960s) in the standard station summaries and analyses however these early data are used to help compare the mean seasonal conditions between Glacier Bay and the greater Gulf of Alaska.

There exists sufficient data to enable comparisons to other long-term environmental records and climate indices; comparisons to these other time series are used to help place the Glacier Bay measurements into a broadly regional context within the greater Northeast Pacific and within a global context with respect to large-scale climate indices. We expect that variability of Glacier Bay's property fields respond to both local processes and regional processes; comparisons to other time series help us identify pertinent linkages.

Methods

Detailed background information, objectives, sampling rationale, sampling design, field procedures, equipment lists, reporting requirements, data handling algorithms are documented in OC2010.1; we provide a brief overview below.

Sampling and Field Protocol

The NPS Glacier Bay CTD makes direct measurements of pressure, temperature, conductivity, chlorophyll fluorescence, dissolved oxygen (DO), photosynthetically available radiation (PAR), and optical backscatter (OBS). Derived values from these measurements yield water column profiles of depth, salinity, density, chlorophyll-a concentration, and turbidity. The sampling plan includes CTD profiles at 1) the "full" station set twice per year: summer (July) and winter (December or January) and 2) the "core" station set in March, April, May, June, August, September, and October. The full station set includes Stations 00–14, 16–21 and 24; the core station set includes Stations 01, 04, 07, 12, 13 16 and 20 (Figure 1). Stations 15, 22 and 23 are no longer sampled. The monthly samples from March through October ensure that the program resolves seasonal signals during the most biologically active period of the year. The higher density sampling in mid-summer and mid-winter ensures that locally isolated ecosystem changes can also be evaluated by this program in the future.

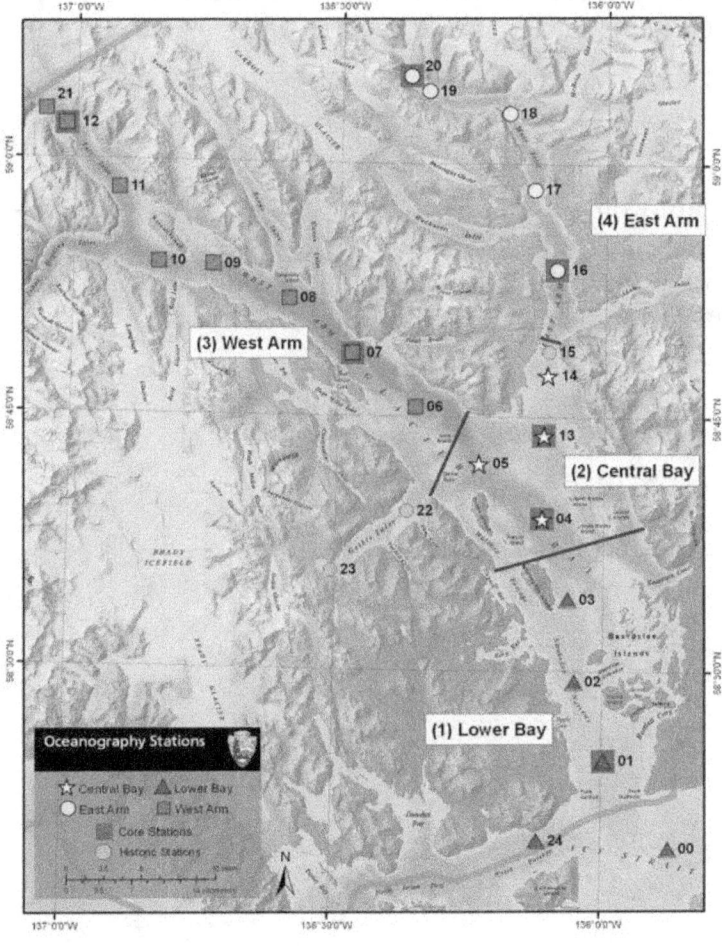

Figure 1. Oceanographic sampling station locations are depicted with numbers and symbols. The four primary domains - separated by solid black lines - are delimited with different station symbols. Stations comprising the core station set are shown on a dark blue square. Blue background shading depicts bathymetric depths, with darker (lighter) colors corresponding to deeper (shallower) waters.

Prior to each downcast, the CTD is allowed to equilibrate just below the surface for two minutes before initiating the vertical profile. The CTD is lowered at 30 m/min in the uppermost 50 m and at 60 m/min for depths below 50m below the surface. Note that samples taken before 2009 employed a shorter surface soak period and did not implement the slower near-surface descent rate.

Processing, Verification and Archiving

The CTD data are processed using the SeaBird, Inc. (2010) SBE Data Processing algorithms as described in OC2010.1. Data are processed to 1 m bins between the surface and the deepest attained depth level. The processing follows the manufacturer recommendations for algorithm settings and is fully automated, including the generation of vertical profile plots of all parameters immediately following a cruise. Visual inspection of the profiles immediately after the cruise by the program manager ensures algorithm and measurement fidelity. Annual calibrations of the CTD at the instrument manufacturer's calibration facility quantifies sensor drift and bounds the accuracy of the measurements. The processed data are either verified for inclusion within the database or additional steps are taken to address problematic profiles as needed.

The oceanographic program protocol (OC2010.1), reports, and data are available at the NPS Southeast Alaska Network (SEAN) I&M website: http://science.nature.nps.gov/im/units/sean/OC_Main.aspx. Original calibration coefficients, raw data and processed data are archived online; verified and final 1-m data are uploaded to the NPS database for permanent storage. Archived material also includes field log sheets and legacy copies of the sampling protocol.

Sampling Coverage

Sampling coverage, delineated by measurement type, year and month, is depicted in figure 2. For field sampling and data processing, the shift from the Hooge et al. (2003) protocol to OC2009.1 was implemented in March 2009. Historic data (1993–2009) were re-processed following OC2010.1. Other oceanographic data for Glacier Bay, which are not part of this database, are archived within the National Ocean Data Center (NODC) World Ocean Database-2009 (WOD09; Boyer et al. 2000).

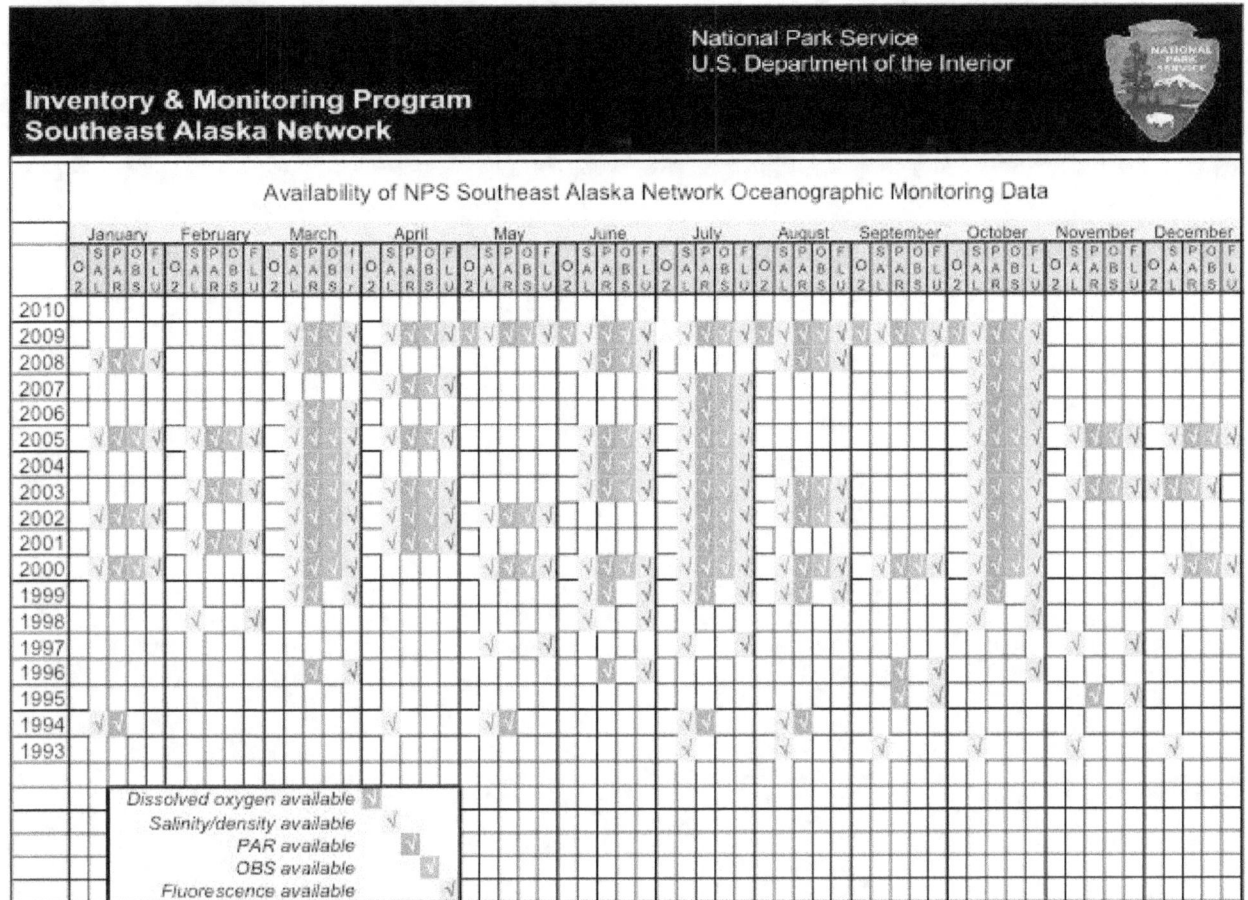

Figure 2. Chart of sampling coverage from 1993 to 2009, depicting sample month and measurement type.

Historic Data Reprocessing

In 2010, the SEAN I&M program undertook a complete re-processing of all historic Glacier Bay oceanographic data collected between 1993 and 2008 following the procedure given in OC2010.1. This reprocessing was warranted because of a number of problems identified in the historic database, including portions of fields appearing in the wrong column, mis-coded latitude/longitude position data and erroneous constant values in measurement fields. As a test case, all data from 2008 were re-processed using the OC2009.1 protocol. Comparing results from the original and re-processed, we found that the OC2009.1 processing resulted in a noticeably cleaner dataset, with fewer T/C/S spikes and fewer density inversions. Additionally, all temperature records from two entire CTD downloads were found to be listed with 0.000 °C readings in the legacy dataset. For these reasons, we concluded that a re-processing of all historical data was warranted. The reprocessing ensures that all data collected since 1993 have been handled similarly to the highest degree possible. Some measurement differences will always exist: early cruises did not employ pumped conductivity cells and a shorter surface soak time, for example. Allowances were made in the algorithms to conform to these special cases. This report represents the first re-analysis of the historical archive with the fully internally consistent reprocessing applied.

Results

Regional Environmental Setting

Understanding the variability represented within the GLBA observations requires insight to local atmospheric and oceanographic processes along with appreciation for external influences, including controlling modes of climate variability and regional weather and oceanographic processes. The climate, weather and oceanography comprise a closely coupled system of interactions and feedbacks and this system also depends on geomorphology, hydrology and cryosphere dynamics. In this section, we examine local, regional and global-scale processes that have some influence upon Glacier Bay oceanographic conditions.

The oceanography of Glacier Bay and the greater Gulf of Alaska is strongly impacted by the position and strength of the Aleutian Low (AL), a pattern of minimum atmospheric pressure at sea level that is centered over the northern North Pacific Ocean for much of the year (Figure 3). The AL reflects the tendency of individual (synoptic) storm systems to propagate from the western North Pacific northeastward into the Bering Sea or eastward into the Gulf of Alaska.

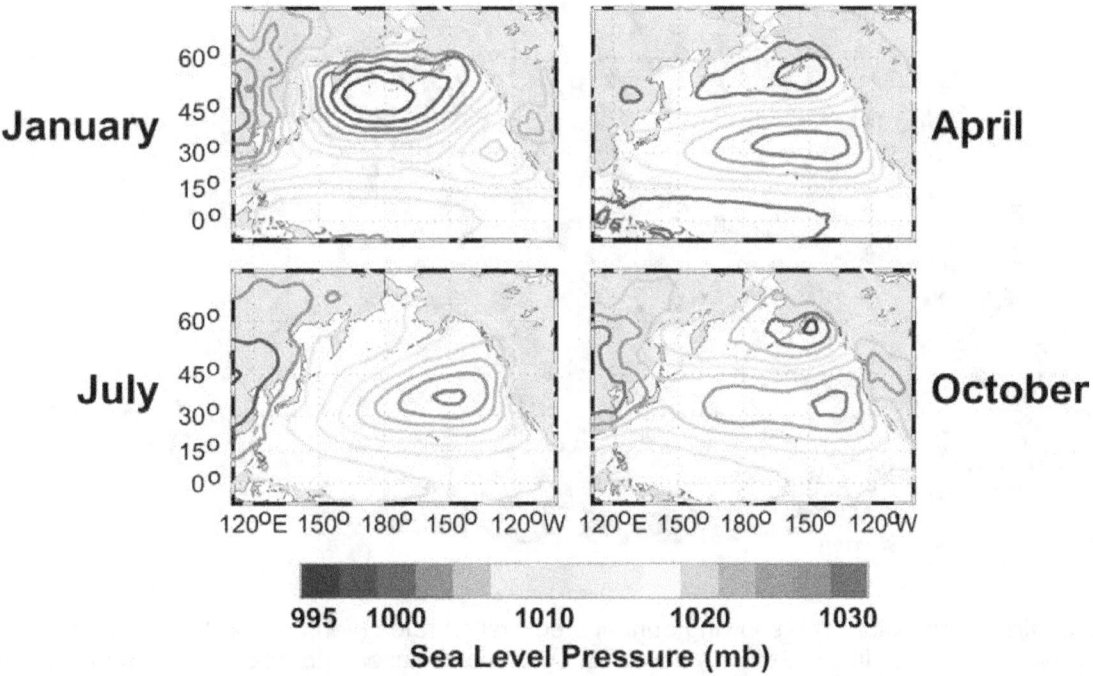

Figure 3. Mean atmospheric pressure at sea level contours for the North Pacific in January, April, July and October. The Aleutian Low (AL) pressure system (blue colors) dominates the eastern Gulf of Alaska in winter, spring and fall months; the North Pacific High (NPH) pressure system (red colors) dominates during summer months.

The sea level pressure (SLP) field varies synoptically, seasonally and inter-annually. On a seasonal basis in the Gulf of Alaska, the North Pacific High (NPH) dominates in summer while the AL dominates in winter. Variability of these patterns directly impact eastern Gulf of Alaska coastal waters. For example, cold continental air (arctic air masses) on a winter storm's northern/western flanks can be advected over the coastal region in mid-winter. Winds associated with AL storm systems' southern flank carry moist air masses from the subtropical Pacific Ocean toward Southeast Alaska, resulting in high precipitation rates and relatively warm temperatures.

Warm conditions also contribute to melting the high altitude snowpack, so a cascade of impacts may follow an eastward displacement and/or strengthening of the AL. For example, temperatures drive the terrestrial melt/freeze cycles, which along with precipitation subsequently determine the volume of coastal fresh water discharge from land. This runoff feeds the Alaska Coastal Current (ACC; Figure 4), a wind and buoyancy driven nearshore current, and helps set the continental shelf density stratification by increasing the nearshore vertical salinity gradient. Strongly stratified coastal waters inhibit vertical mixing and so reduce entrainment of sub-pycnocline nutrients into the euphotic zone where primary producers are dependent upon nutrients for growth and reproduction. In this fashion the AL indirectly impacts the continental shelf density field; direct impact from the AL comes in the form of wind stress, which provides energy for mixing (breakdown of density stratification) and wind-driven advection.

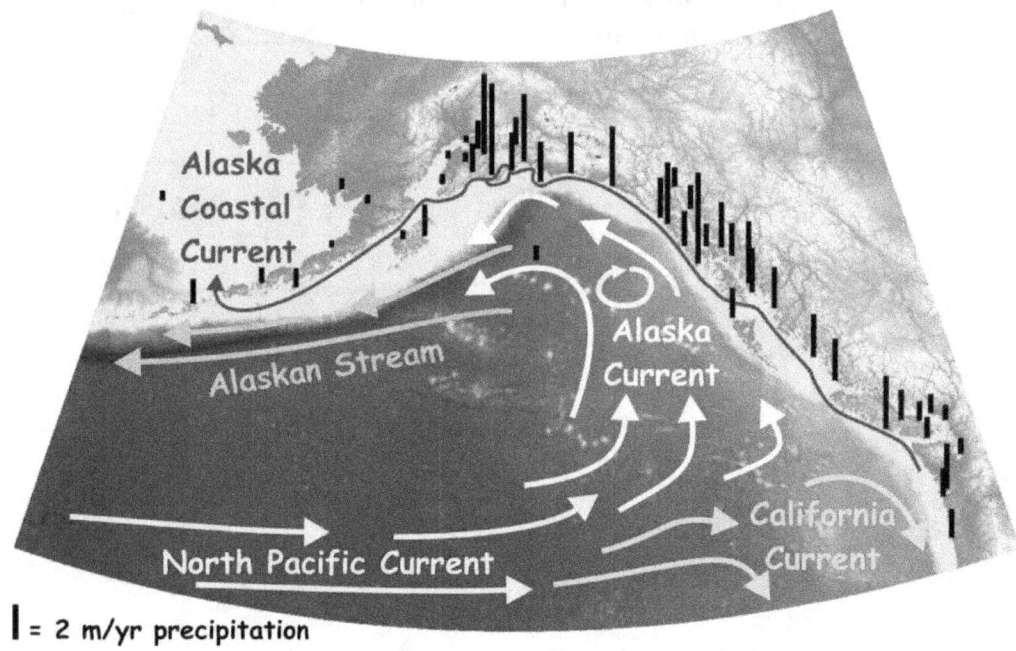

Figure 4. Idealized circulation map showing annual precipitation rates (vertical bars) and the large-scale oceanic circulation features in the Gulf of Alaska. Topographic elevations are shaded in green to brown; bathymetric depths are shaded in blue. Reprinted from Weingartner et al. (2005).

The annual progression of earth's seasons modulates the amount of incoming solar radiation; heat and light are first-order controls upon surface air temperatures, the biosphere's growing season and ocean-atmosphere heat exchanges. A monthly climatology of the annual temperature cycle recorded at Juneau, Alaska is shown in figure 5 along with companion panels showing SLP, wind speed and precipitation. During summer months, the ocean gains heat directly from incoming solar radiation; during winter months, the ocean loses heat to the atmosphere through cooling from long-wave radiation heat loss. The maximum air temperature (13.8°C) occurs in July and is in phase with the summer SLP maximum. However, the minimum (-2.3°C) in January lags the SLP fall minimum, showing that the annual cycles of heating/cooling and storminess do not progress in tight synchrony with each other. Likewise, the oceanic temperatures do not share identical timing with the air temperatures: the maximum near-surface water temperature at Station 04 occurs in July-August and the minimum occurs in March-April. At 200 m depth, maximum water temperatures are observed in November; minimum temperatures are in phase with those at the surface.

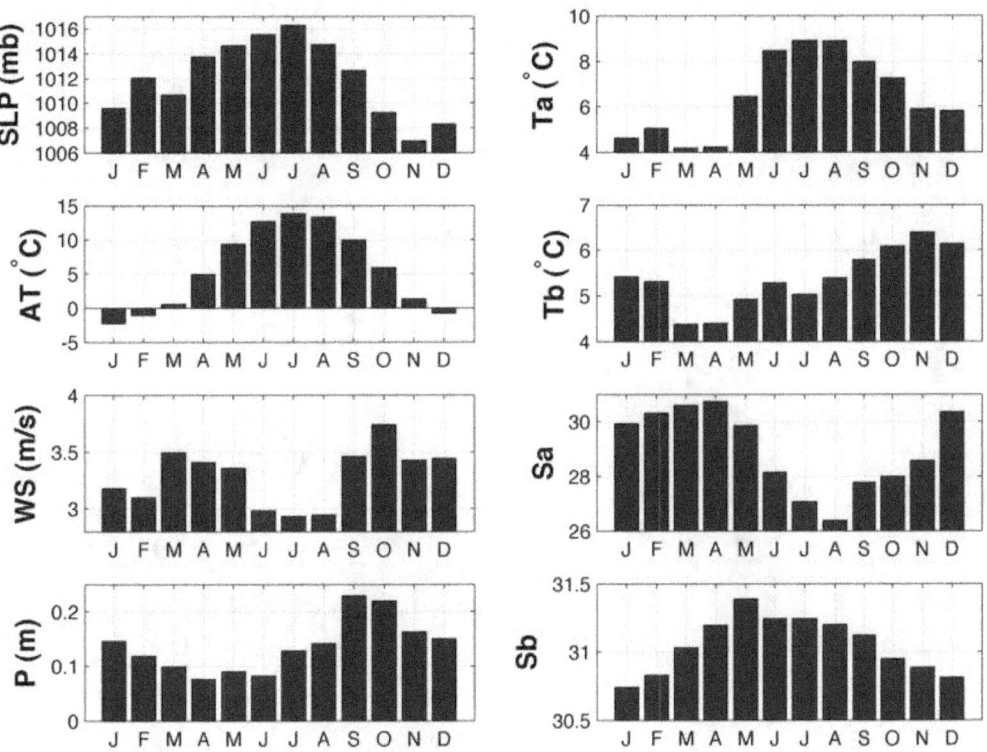

Figure 5. Monthly climatology of atmospheric conditions recorded at Juneau, Alaska over 1992–2009 (left column) and mean monthly T and S values averaged over 0–10 m (Ta, Sa) and 195–205 m (Tb, Sb) depth strata at Glacier Bay Station 04 (right column). Atmospheric variables include sea level pressure (SLP), air temperature (AT), wind speed (WS) and precipitation (P). Station 04 T and S values in some months may be biased due to low numbers of samples.

Precipitation peaks in early fall (September) and has its annual minimum in late spring and early summer (April-June). Thus, coastal discharge is reduced during the winter months due to mean freezing temperatures between December and February and low precipitation rates following the end of winter. In contrast, near-surface salinities at Station 04 are at their annual minimum prior to the peak in precipitation; this demonstrates the importance of snowmelt to the marine fresh water budget. Salinities increase through the winter: maximum salinity occurs in April, near the time of minimum water temperature. At 200 m depth, maximum salinities occur in May (shortly after the temperature minimum); minimum salinity at depth occurs in January.

Wind speeds exhibit a bi-modal structure with peaks in the fall (October) and the spring (March), which induces strong mechanical mixing of the near-surface waters. The oceanic cycle of stratification offers varying levels of resistance to mixing between these two maxima: October closely follows the period of maximum water column stratification of late summer while March represents the time of minimum stratification because it follows strong winter wind mixing, convective cooling and reduced levels of coastal discharge. Figure 6 shows the magnitude and sign of the downwelling wind stress in the northeastern Gulf of Alaska. This component of the wind forcing helps drive the ACC because strong downwelling winds help keep the fresh ACC plume close to shore. By keeping the plume close to shore, the oceanic horizontal (cross-shore) density gradient increases, thereby accelerating the ACC in the along-shore direction. Thus, advective connections (ACC strength and location) along the eastern Gulf of Alaska coastal zone are also modulated by the wind field, which is set up by the position and strength of the AL's storm systems.

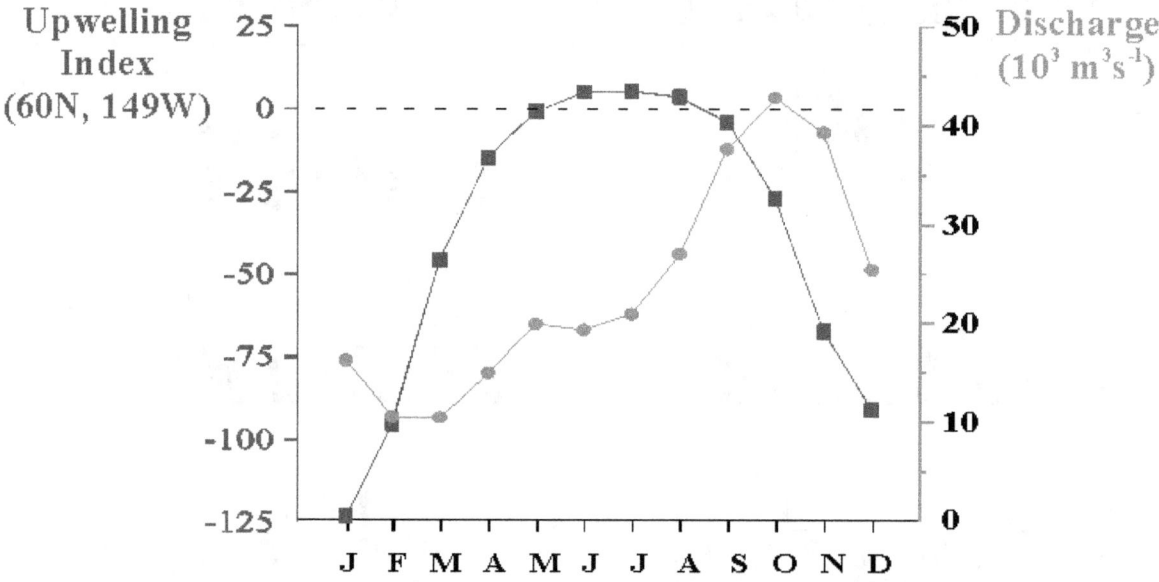

Figure 6. Annual cycle of the northeastern Gulf of Alaska Upwelling Index (left) and the coastal fresh water discharge (right), after Royer (1982). Note that the discharge cycle is representative of the entire northeastern Gulf of Alaska; timing of local discharge in Glacier Bay likely varies somewhat.

Together, the panels of figures 5 and 6 show that the relative timings of the atmospheric and terrestrial inputs are only loosely coupled to oceanic conditions, leading us to conclude that internal ocean dynamics and other external (local or regional) atmospheric processes also must play a role in setting the seasonal character of Glacier Bay waters.

The Glacier Bay oceanographic time series extends from 1993 to present so we compile monthly anomalies of the SLP, temperature, wind and precipitation records over this time period (Figure 7). Anomalies in figure 7 scaled ensure zero mean; units of the measurement parameters are retained. The sea level pressure record over this time frame is incomplete. A five-month smoothing applied to these data show that some low-frequency fluctuations lie behind the dominant high-frequency variability. For example, the period from 1996 to 2002 exhibits generally higher wind speeds than the period of 2002–2010. There are no trends evident in the temperature record although three months stand out with large negative anomalies in 1993, 1995 and 2006.

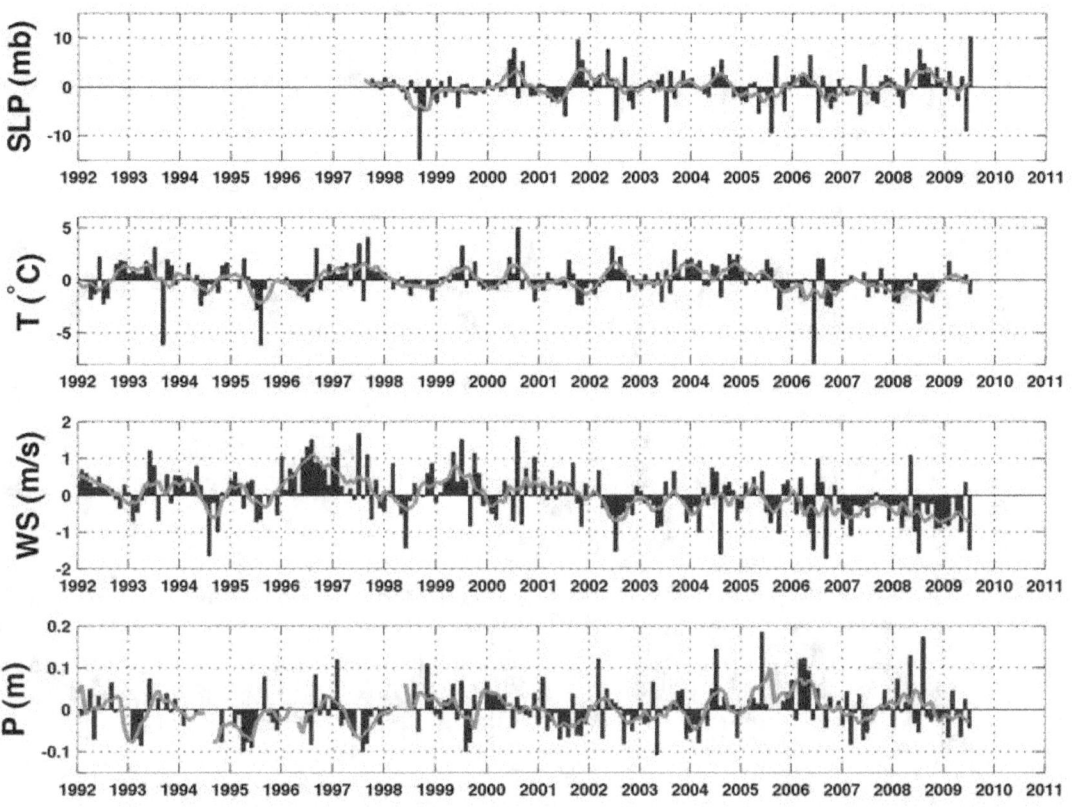

Figure 7. Monthly anomalies of sea level pressure (SLP), temperature (T), wind speed (WS), and precipitation (P) recorded at Juneau (PAJN) over 1992–2009. Red lines depict a five-month moving average.

The local conditions recorded at the PAJN station are modulated by the global climate, which evolves under the influence of numerous modes of atmospheric and oceanographic processes. Major climate indices that track some of these fluctuations are depicted in figure 8: the Pacific Decadal Oscillation (PDO) sea surface temperature index, the North Pacific Gyre Oscillation (NPGO) sea level height index, the El Niño-Southern Oscillation (ENSO) sea surface temperature index, and the Pacific-North American (PNA) sea level pressure index. These indices, shown over 1950 to present, are plotted in figure 8. They are shown as normalized anomalies: each record is transformed by removing the monthly mean value and dividing by the standard deviation so that each resultant index has a mean of zero, unity standard deviation, and no associated units.

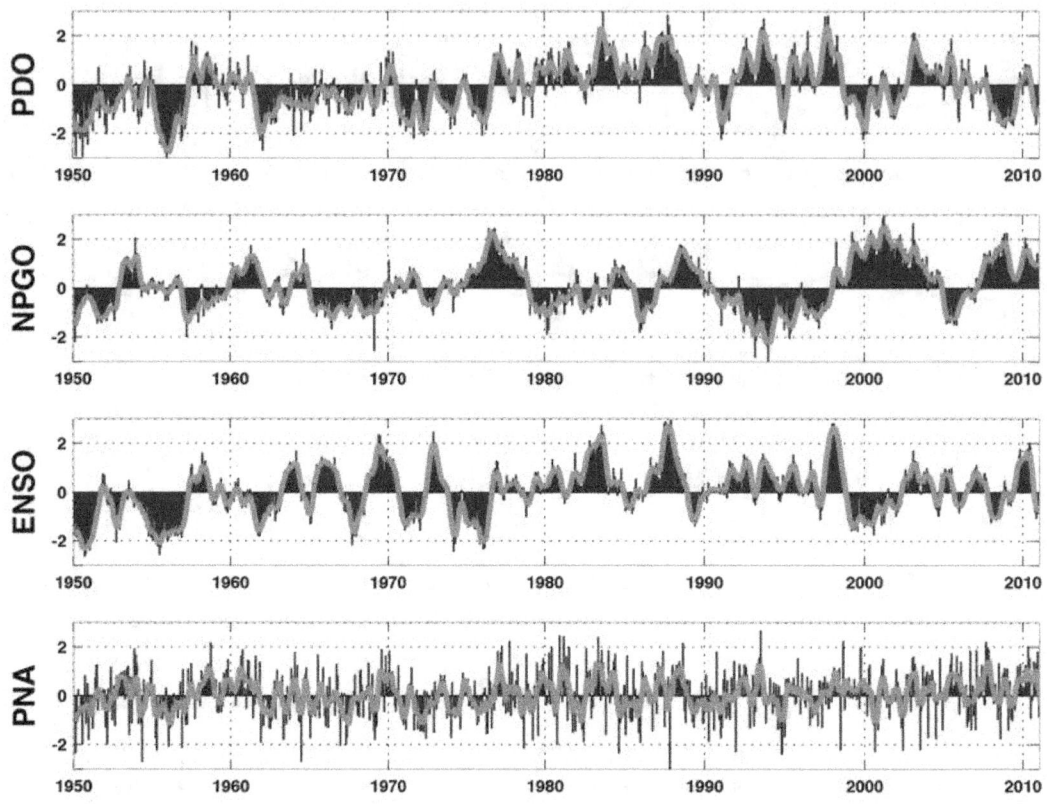

Figure 8. Normalized monthly anomalies of the following major climate indices: the Pacific Decadal Oscillation (PDO) sea surface temperature index, the North Pacific Gyre Oscillation (NPGO) sea level height index, the El Niño-Southern Oscillation (ENSO) sea surface temperature index, and the Pacific-North American (PNA) sea level pressure index. Red lines depict a five-month moving average for each record.

The PDO is the leading mode of North Pacific sea surface temperature (SST) variability (Mantua et. al, 1997). It exhibits decadal-scale and multi-decadal scale temporal variability and its spatial pattern shows that the western North Pacific fluctuates in an out-of-phase relation to the coastal Gulf of Alaska and the eastern equatorial Pacific (Figure 9).

ENSO is a tropical climate mode of variability with dominant fluctuations in the 3–5 year time frame. Development of El-Niño conditions triggers a northward-propagating Kelvin wave along the eastern North Pacific and this wave results in oceanic temperature fluctuations observed from the equator to the northern Gulf of Alaska. While the ENSO and PDO patterns are similar in shape, the locations where the largest amplitude fluctuations occur are different, their spectral contents differ, and the associated wind stress patterns differ.

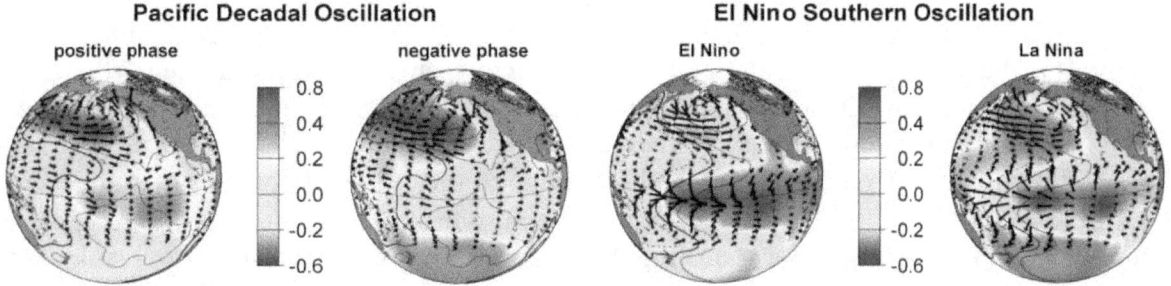

Figure 9. Structure of the PDO (left two panels) and ENSO (right two panels) modes of SST variability. Colors depict the SST anomaly, vectors depict wind stress anomalies and contours denote structure of the SLP field. Images from http://jisao.washington.edu/pdo/.

The NPGO is the second dominant pattern of sea surface height variability in the North Pacific Ocean (Figure 10) and its spatial structure shows a three-pole pattern in the eastern North Pacific, with the Gulf of Alaska and the region west of California and Mexico fluctuating in phase with each other and an out-of-phase center situated between the two (DiLorenzo et al. 2008). The NPGO is closely correlated with salinity and nitrate fluctuations along the southern Gulf of Alaska Line P section (Figure 10). Nitrate is an important macronutrient that often limits marine primary productivity in the North Pacific.

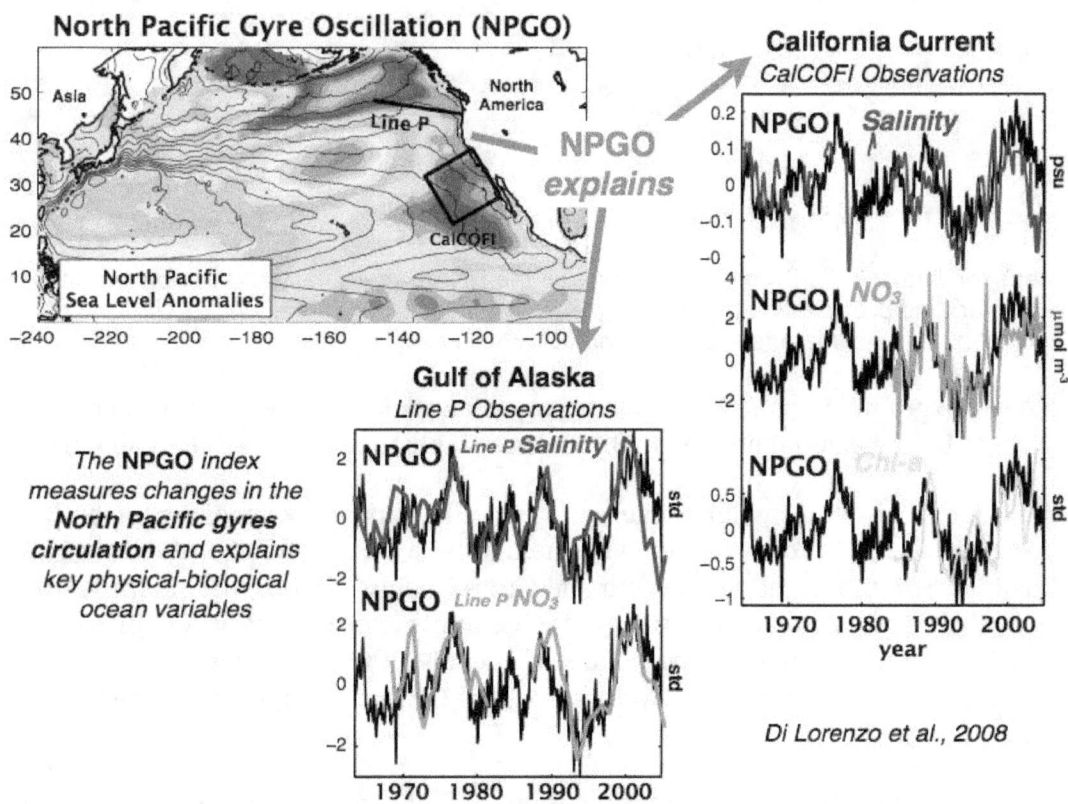

Figure 10. Spatial pattern of the NPGO (upper left) along with time series of salinity, nitrate and chlorophyll fluorescence fluctuations along Line P (lower) and in the CalCOFI observation area (right). Image from http://www.o3d.org/npgo/.

Whereas the PDO, the NPGO and ENSO are all measures of oceanic variability (with corresponding atmospheric signals), the PNA (Figure 11) reflects atmospheric conditions more directly; it is the second leading mode of atmospheric pressure variability in the northern hemisphere, normally evaluated at the 700 mbar level (van den Dool et al. 2000). The PNA exhibits much higher frequency variability than the oceanic indices. The PNA is closely related to the North Pacific (NP) index, which is the average of the sea level pressure over the box delimited by (30–65°N, 160°E–140°W) and so is a measure of the strength of the AL pressure system over its normal winter location (Figure 3). The PNA's pattern depicts opposing phases centered over the northern North Pacific and the continental United States (Figure 11). The location and strength of AL storm systems is the North Pacific manifestation of the PNA and so it controls or impacts advection of air masses, cloud cover, precipitation rates and air temperatures, along with oceanic wind-driven mixing and wind-driven circulation fields.

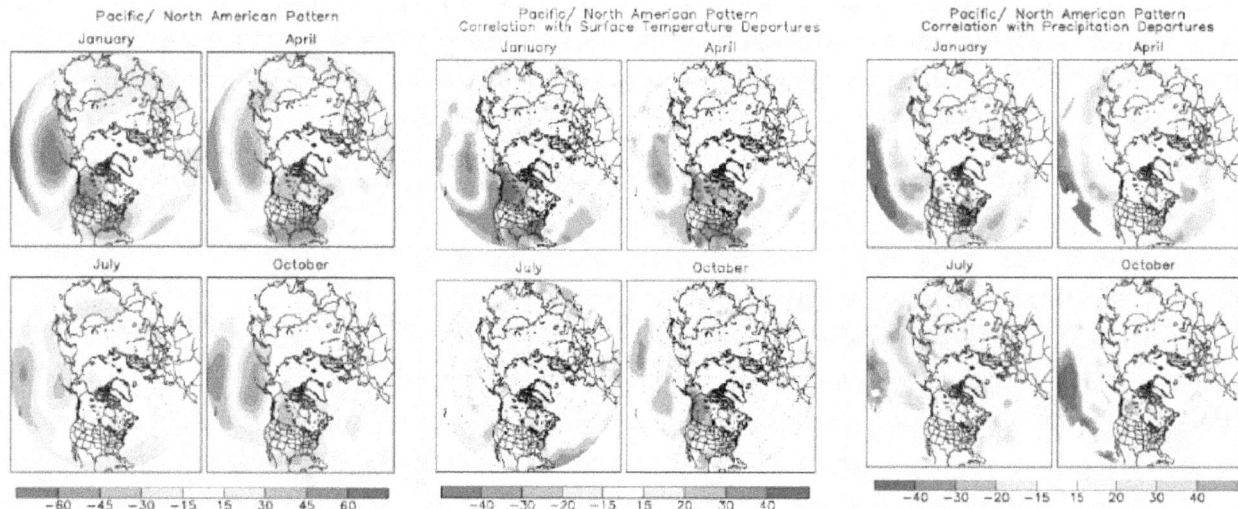

Figure 11. Maps for January, April, July and October of the PNA pattern (left) and its correlation with temperature (center) and precipitation (right). Images from http://www.cpc.ncep.noaa.gov/data/teledoc/pna.shtml.

With the above background in hand, we now turn to oceanographic observations to complete our regional overview before examining the GLBA oceanographic monitoring data in detail.

While oceanographic conditions within Glacier Bay are likely strongly dependent upon local weather and dynamics, the fjord is coupled with the marine waters of Southeast Alaska: Icy Strait and Chatham Strait to the east and south and Cross Sound and the Gulf of Alaska to the west. Shallow subsurface sills and the narrow entrance limits exchange of waters into and out of the fjord. Based on numerical model experiments, Hill et al. (2009) conclude that waters of Glacier Bay primarily communicate with waters from the Cross Sound region to the west, rather than from the east.

Using data from the WOD09 archive, we form a transect that extends from the central Gulf of Alaska to the northern reaches of Glacier Bay. The panels in figure 12 and figure 13 show an example of averaged observations between these two systems in late winter/early spring and in late summer/early fall. Note that these plots are generated from non-synoptic data spanning multiple years so care must be taken in their interpretation. Nevertheless, seasonal cycles likely dominate over inter-annual variability to a large extent so these sections are probably at least representative of synoptic snapshots. The largest apparent discontinuity is observed in the thermal field, which shows Glacier Bay temperatures at the end of winter to be appreciably colder north of the main sill than to the south. While this feature may reflect actual springtime conditions, it may also be a result of temporal aliasing.

Figure 12 shows that late winter waters are nearly homogeneous within the fjord, having been exposed to months of reduced runoff (due to accumulation of the snow pack field through the winter), convective cooling and vigorous wind and topographically-induced mixing. Water temperatures are cold (2.5–4.5°C) and salinities vary little at depth (S~31) although a shallow (<25 m thick) low-salinity plume can extend south from the head of the fjord even at the end of winter. Offshore waters are weakly stratified ($\Delta T<3°C$; $\Delta S<1.5$) over the upper 300 m at this time of year with most of the stratification occurring near ~150m depth.

In contrast, the end-of-summer thermal gradient over the upper 300m of the central Gulf water column increases to $\Delta T\sim 10°C$ (the salinity gradient increases slightly to $\Delta S\sim 2$) and most of the stratification occurs in the upper 50 m of the water column (Figure 13). Inside the fjord at the end of summer, the water is only moderately thermally stratified ($\Delta T<4°C$) but the salinity stratification can be extremely large in the upper 50 m ($\Delta S\sim 5$–20).

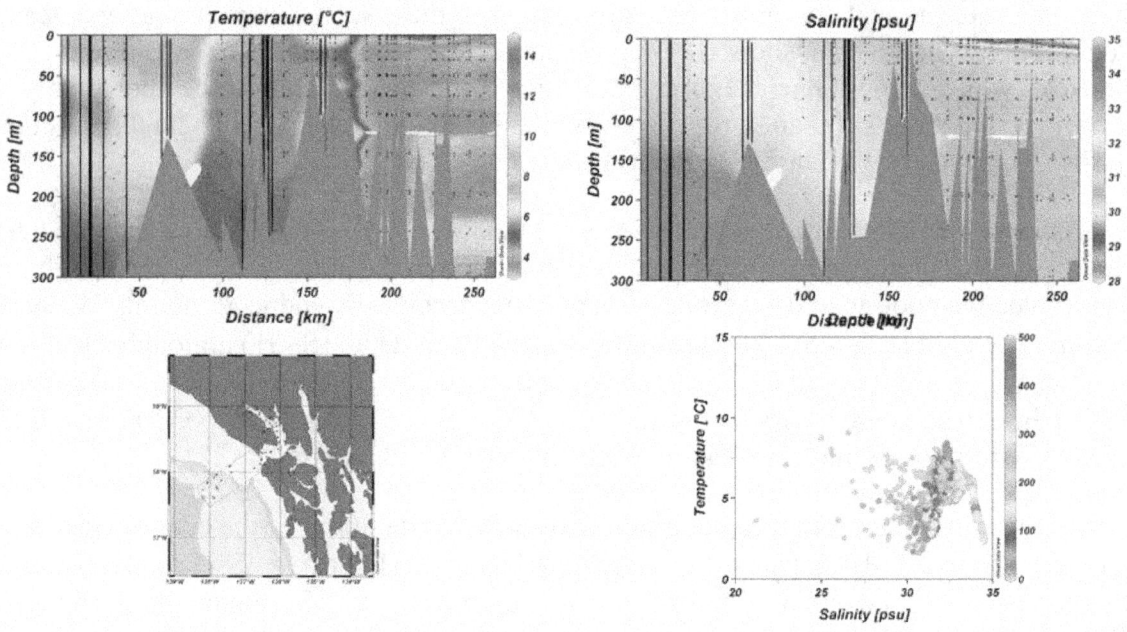

Figure 12. Late winter/early spring ocean conditions (March to May) between the Gulf of Alaska and the northwest reaches of Glacier Bay. The cross-sections are oriented with the farthest offshore sites at 0 km and upper bay fjord stations beyond 250 km. Note the separation of the fjord stations from the offshore stations on the T-S diagram. Black dots and vertical lines on the cross-sections denote locations of data used in the figure. T-S diagram points are colored by depth (m). Data are extracted from Boyer et al. (2009).

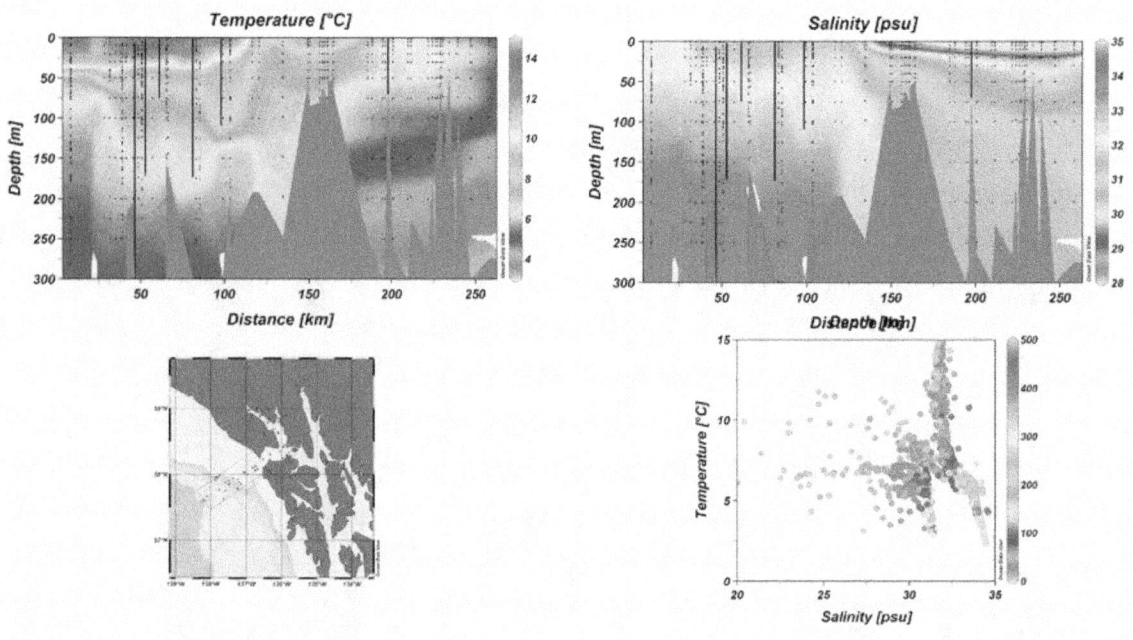

Figure 13. Late summer/early fall ocean conditions (August to September) between the Gulf of Alaska and the northwest reaches of Glacier Bay. See Figure 12 caption for details.

Mean Monthly Conditions, Anomalies and Trends

Previous studies (e.g., Etherington et al. 2007) identified four primary biophysical domains within Glacier Bay (Figure 1):

1) Lower Bay stations are those in Sitakaday Narrows and Icy Strait (00–03 and 24).
2) Central Bay stations (04, 05, and 13–15) are those north of the main sill at the widest portion of the fjord.
3) East Arm stations (16–20).
4) West Arm stations (06–12 and 21).

In this section, we examine one station from each domain in detail (Stations 01, 04, 12 and 20); companion figures for all stations are included within the appendices. Because the upper ocean typically experiences strong dynamical forcing due to the presence of the ocean-atmosphere interface and the associated stresses and fluxes at this interface, our focus will primarily be on the upper 50 m for the Lower Bay, West Arm and East Arm stations. We expand the focus to the entire water column for the Central Bay station.

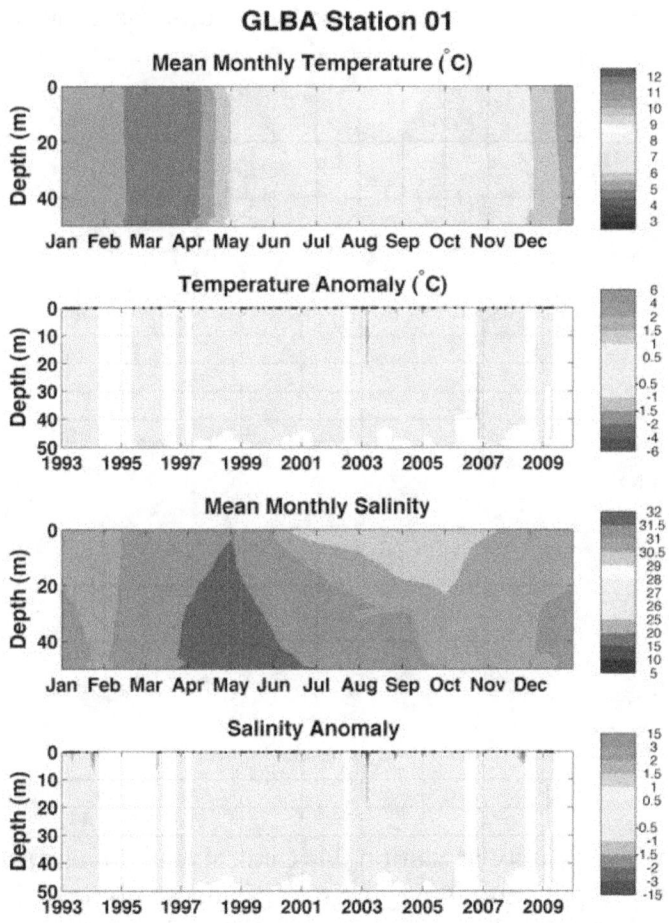

Figure 14. Station 01 cross-sections of the mean monthly temperature and salinity fields (first and third rows, respectively) along with companion depth-time monthly anomaly plots (second and fourth rows, respectively).

Station 01 (Lower Bay domain) lies within the shallow, strongly mixed regime near the main (southern) sill and so shows only weak levels of stratification even through the summer months. Because of the strong mixing at this site, the phasing of maximum and minimum temperature and salinity near the surface and near the bottom are nearly in phase with one another. Mean monthly surface temperatures are not as high as at the stratified stations farther north in the fjord. Salinities at Station 01 are greatest near the bottom in May, having increased from February, supporting the notion that deepwater renewal may occur primarily in winter and spring. Even at depths exceeding 200 m, mean salinities at stations in the other three domains do not attain the highest values observed at the (shallow) Lower Bay sites, where mean salinities can exceed 31.5, so waters must experience some degree of mixing as they descend to their equilibrium depth.

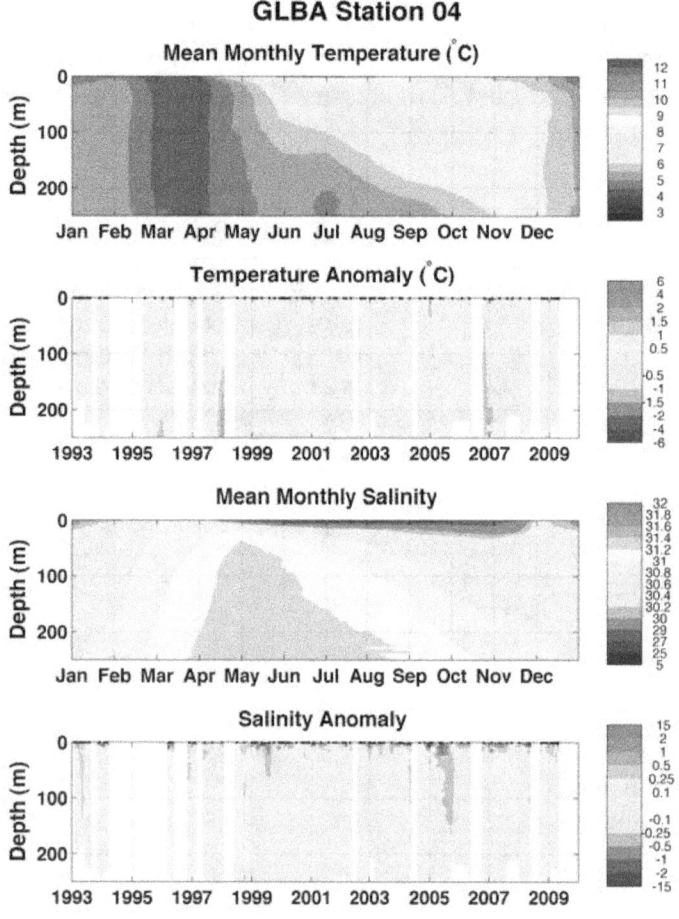

Figure 15. Station 04 cross-sections of the mean monthly temperature and salinity fields (first and third rows, respectively) along with companion depth-time anomaly plots (second and fourth rows, respectively).

The mean annual cycles of temperature and salinity at Station 04 (Central Bay domain) show that waters near the surface and waters at depth exhibit varying phases and amplitudes. For example, maximum temperature occurs in July at the surface, but at 200 m depth the maximum temperature occurs in November. Minimum temperatures across the water column occur in March. Maximum (minimum) salinities occur in March (October) at the surface and maximum (minimum) values occur in May (December) at 200 m depth. In aggregate, these relations suggest that the upper and lower portions of the water column are responding to the annual cycles of two different forcing mechanisms (refer also to Figure 5).

Figure 16 shows mean monthly temperature, salinity and surface-referenced density (sigma-0) at Stations 01 (the shallowest station north of the fjord mouth) and Station 04 (the southernmost deep station). Highest densities at all depth at Station 04 occur in May. Subsurface salinity at Station 04 increases into May (the surface salinity maximum occurs earlier, see Figure 15). The annual minimum in density occurs first at 45 m depth and 100 m depths at Stations 01 and 04 in October and November and then is followed in January at the 250 m depth at Station 04. The surface density at Station 04 is always less dense than that at Station 00 (not shown). Monthly mean densities in January, March and April above the sill achieve greater density than waters at much greater depths within the fjord (note the blue trace in the bottom panel of Figure 16 crosses the black and gray traces).

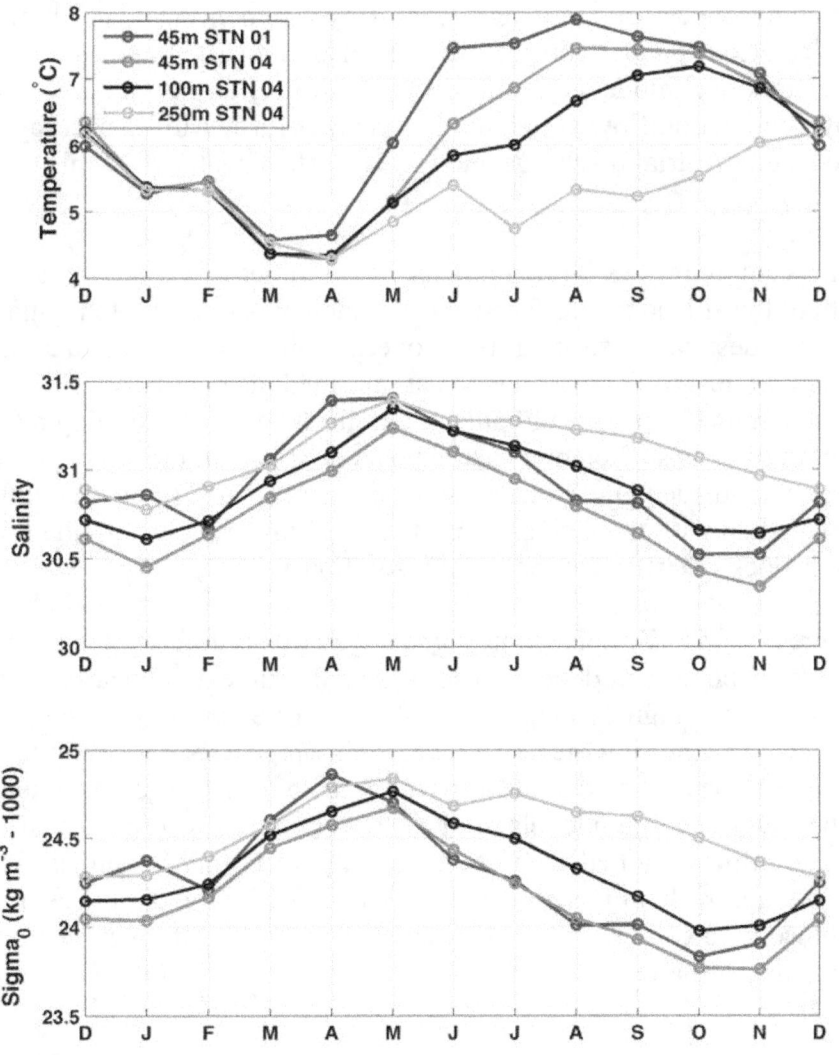

Figure 16. Mean monthly time series of temperature (top), salinity (middle) and sigma-0 (bottom) for 45 m (red and blue), 100 m (black) and 250 m (gray) depth strata at Stations 01 and 04.

The timing of deepwater renewal has been an issue of some discussion in the literature (see second paragraph of the Introduction). Average density of the upper 40 m at Station 00 typically falls below the 100 m and 250 m density at Station 04 in May and remains less dense until October or November, in close agreement with Matthews' (1981) assessment. These observations provide evidence that the fjord's deepwater renewal is on average primarily limited to one (extended) portion of the year; because the waters are not in hydrostatic balance here we expect that there is some degree of hydraulic control on the intrusion of dense water. The analysis here does not consider the impact of interannual variability; it is possible that some years do promote flushing of deep depths during summer months as well.

Valle-Levinson and Wilson (1994) show that strong barotropic forcing over a sill leads to asymmetry of the advection of the gradient field and that the tidal prism mediates cross-sill exchange, rather than the primarily hydraulic control expected in fjords with weaker forcing. The model of Hill (2007) exhibits excursions on the order of 20 km in the lower bay so advection of density by the tides may well introduce Icy Strait and Lower Bay waters into and north of the sill region where gravity flows can exert greater control. We note also that Hill et al. (2009) analyzed the currents of a barotropic numerical model and found net northward flux into the bay in the center of Sitakaday Narrows and net outflow at the margins. This is further influenced by the net outflow required to balance the terrestrial discharge and the subsurface waters that this transport entrains.

Near surface (0–40 m) waters south of the sill remain denser at all times of year than the corresponding waters north of the sill, so the surface waters tendency is to promote regular estuarine exchange year round, despite the strong mixing over the sill. It is possible that inflow during summer and fall months (June to October) occurs at intermediate depths. While an upwelling or a Bernoulli suction mechanism could pull waters from below the sill depth (perhaps on each tidal cycle), we currently lack sufficient measurements deeper than the sill depth in Icy Strait and so are unable to begin an adequate assessment these mechanisms. (Note that this problem will be partially remedied with the inclusion of Station 24 into the core station set in 2011. Another problem is the lack of sufficient temporal resolution.)

The anomaly panels of figures 15 and 17 depict some signals that are confined to the near-surface; some signals that are found only at depth and some signals that extend nearly uniformly across the water column. These latter signals suggest the influence of external forcing upon the Glacier Bay system because waters entering the bay are nearly homogenized by strong mixing in Sitakaday Narrows and because surface and deepwaters within the bay are largely isolated from one another for much of the year due to high levels of upper water column vertical stratification. Near-surface temperature and salinity anomalies are likely locally phenomena resultant from high/low rates of heating/cooling, freshening/salinization associated with the cloud cover anomalies (and associated incoming solar radiation anomalies) and precipitation anomalies (and subsequent terrestrial discharge anomalies).

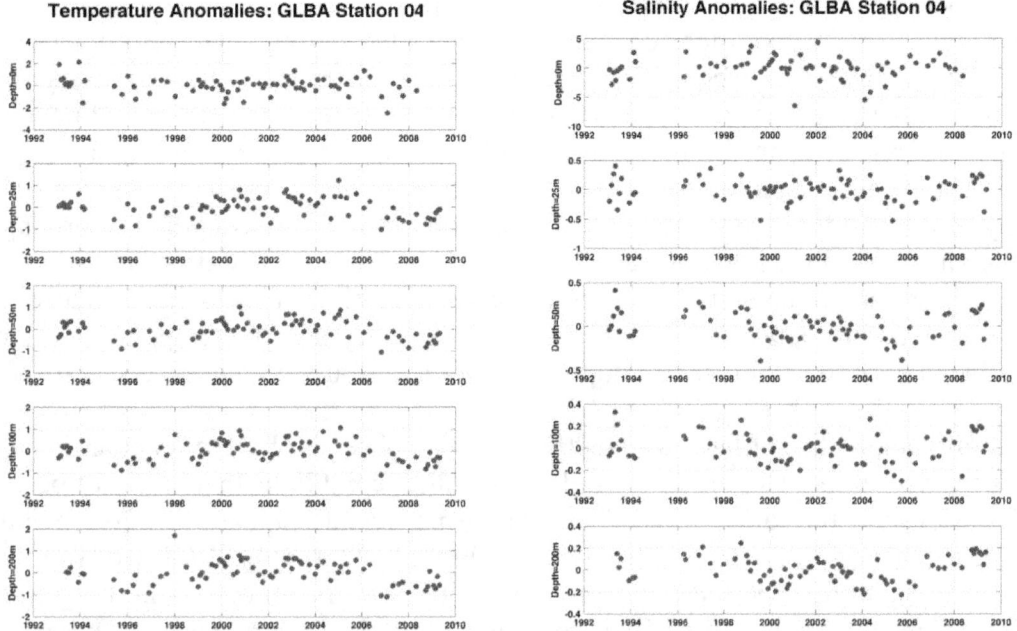

Figure 17. Time series of temperature and salinity anomalies at select depths. Note change of vertical scale at different depths.

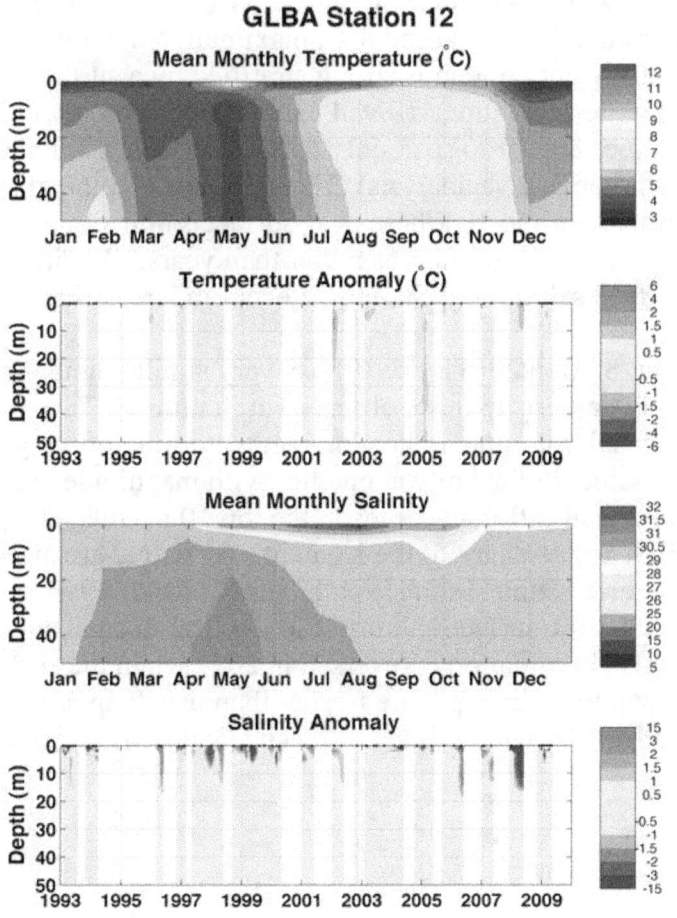

Figure 18. Station 12 upper 50 m cross-sections of the mean monthly temperature and salinity fields (first and third rows respectively) along with companion depth-time monthly anomaly plots (second and fourth rows respectively).

The strong temperature anomaly observed at depth in late 1997 and early 1998 is the footprint of the record-strength El Niño conditions of that year. The El Niño triggers a coastally-trapped disturbance (Kelvin wave) that propagates along the North American west coast on its right-hand side (due to the influence of the earth's rotation). This wave displaces the thermocline, resulting in a temporary increase of temperatures as the trough of the wave passes. The signal is barotropic (operates across the entire depth of the water column) but the surface waters are more strongly influenced by local conditions so the wave is strongly observed only at depth. The propagation time for this signal to reach Alaskan waters is on the order of half a year after the El Niño disturbance impinges upon the eastern equatorial Pacific shoreline. This signal was also observed at depth at oceanographic station GAK1 and Royer (2005) attributes the temperature increase to anomalous advection of the along-shelf temperature gradient found in the eastern Gulf of Alaska.

The cold and salty anomalies of late 2006–2009 were also observed by Janout et al. (2010) and represent a (possibly temporary) return to colder and saltier conditions that have not been seen in the Gulf of Alaska since the early 1970s. This signal is particularly apparent at 200 m depth where a prolonged period of fresh conditions from 1999 to 2006 (Figure 17, lower-right panel) sets off the subsequent saltier period. The shift from cold to warm conditions in the greater Gulf of Alaska were coincident with a large-scale reorganization of the marine ecosystem: crab and shrimp stocks crashed in the late 1970s and the fishery was replaced by a pelagic fishery (Anderson and Piatt 1999).

Located at the head of the West Arm, the 0–50 m observations at Station 12 (Figure 18) show that, near the glacial terminus, mean temperatures never exceed 8°C, maximum temperatures normally occur at sub-surface depths, and the water column is strong stratified by a surface fresh water lens between May and October. The largest and longest-lived temperature anomaly occurs over at least half of 2005; most other anomalies appear to span only portions of a year (although there are many data gaps that preclude a more definitive analysis). This suggests that the glacial melt, which is presumably responsible for many or most of these near-surface salinity anomalies, fluctuates with a characteristic time scale on the order of months rather than years. As with stations farther south, the maximum subsurface salinities at Station 12 occur in late spring.

Located at the northern end of the East Arm, Station 20 (Figure 19) has very similar annual cycles, strength of stratification and anomaly patterns to those observed at Station 12. The salinity anomalies sometimes alternate fresh/salty in time, indicating that the timing of the glacial discharge varies in different years. Some instances of salinity anomalies with magnitude greater than +/-1 exhibit opposite-sign anomalies lying at different depths in the top 10 m, suggesting that plume dynamics which govern the depth and strength of the ice melt pycnocline are affected by mixing with deeper, more saline water. For example, in the first three samples of 1999, a fresh anomaly lies over a more saline anomaly: the surface salinity is lower than normal and just below this the waters are saltier than normal. The situation is reversed in 2001 when the surface waters are anomalously salty but waters down to 25 m depth are fresher than usual; apparently the ice melt has been mixed downward, perhaps by wind mixing or mechanical stirring from the presence of deep ice keels.

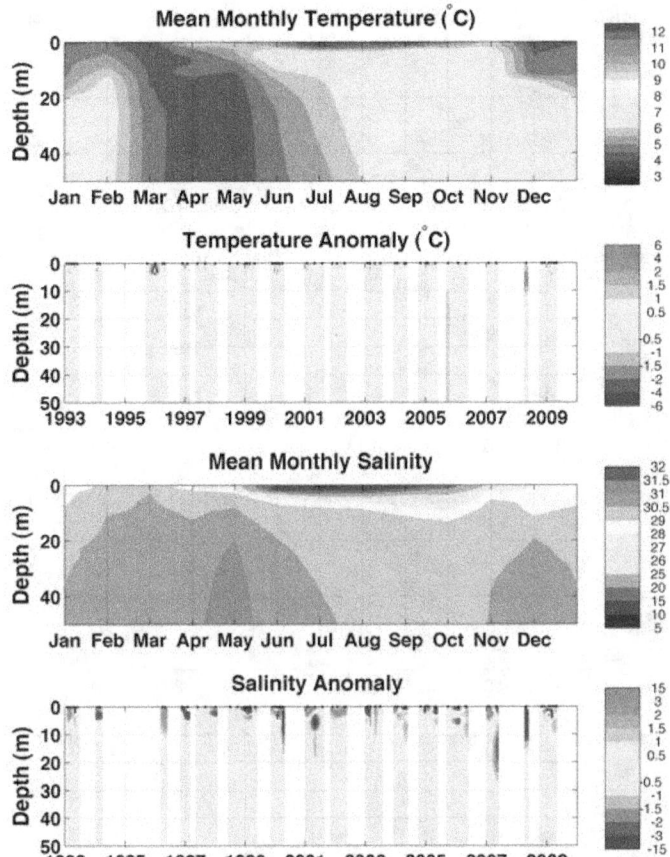

Figure 19. Station 20 upper 50 m cross-sections of the mean monthly temperature and salinity fields (first and third rows respectively) along with companion depth-time monthly anomaly plots (second and fourth rows respectively).

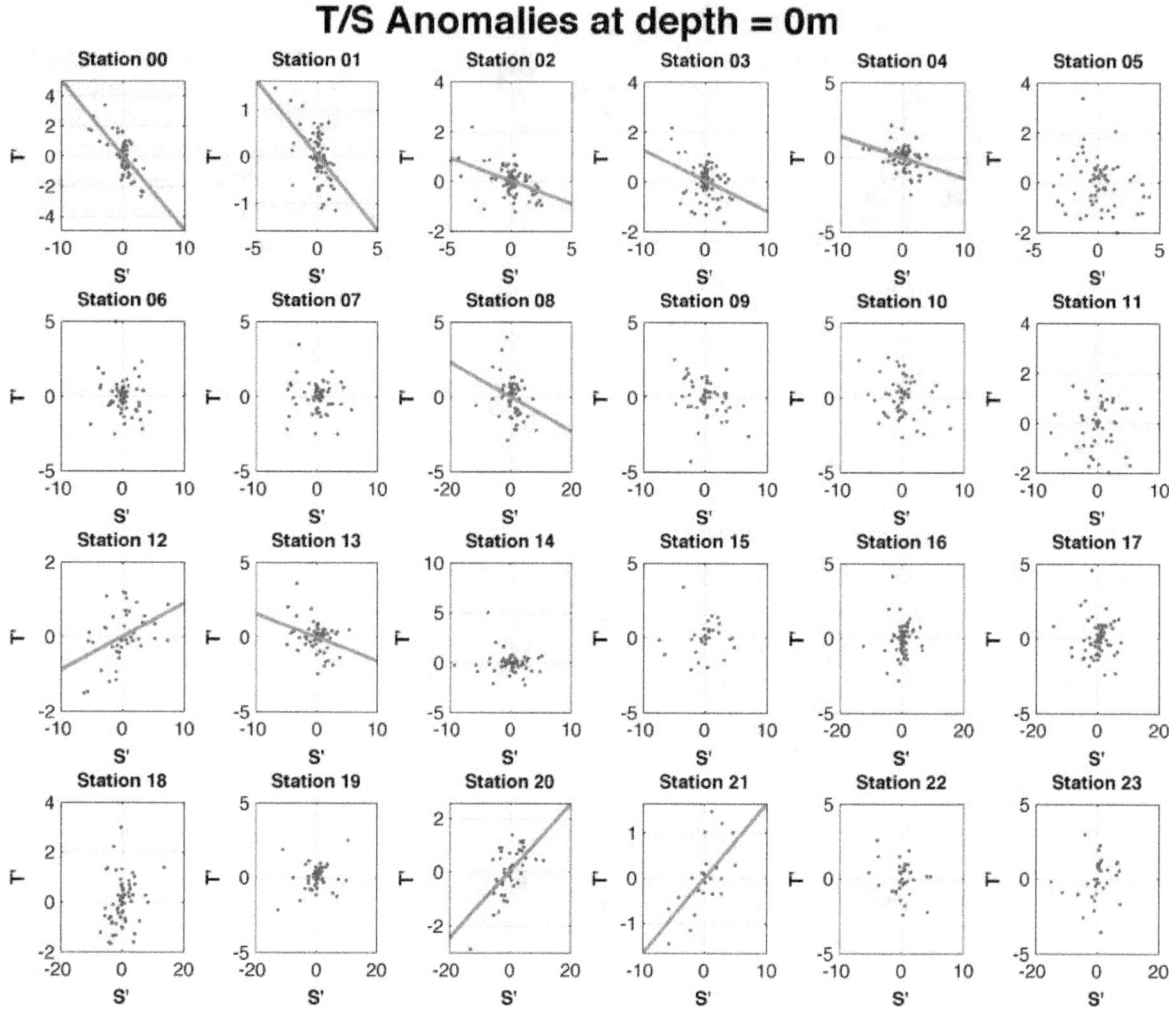

Figure 20. Anomalies of temperature (T') vs. anomalies of salinity (S') at 0 m depth for all stations. Significant relations at the 95% level are shown with a red best-fit linear trend line. Note that each panel is individually scaled to fit the range of observed anomalies.

We examined the relation of temperature and salinity anomalies with respect to each other at all stations for surface (0 m; Figure 20) and subsurface (100 m; Figure 21) depths. At 100 m depth all stations north of the Lower Bay domain show a statistically significant negative relation between T and S: warm and fresh anomalies occur together and cold and salty anomalies occur together. This relation is consistent with our expectations that warm anomalies are associated with enhanced atmospheric advection of warm, moist air to the region which results in higher rates of precipitation, snow pack melt and coastal runoff.

The negative T/S slope is also found at all Lower Bay stations (measurements near the surface) and at Stations 06 and 13. Most other near-surface measurements show no significant relation although Stations 12, 20 and 21 (stations closest to the head of the West Arm and East Arm fjords) depict a positive relation. This suggests that near the glaciers the T/S dynamics are instead controlled by glacial melt: cold (warm) and fresh (salty) anomalies are coincident (melting ice releases cold, fresh water and vice-versa).

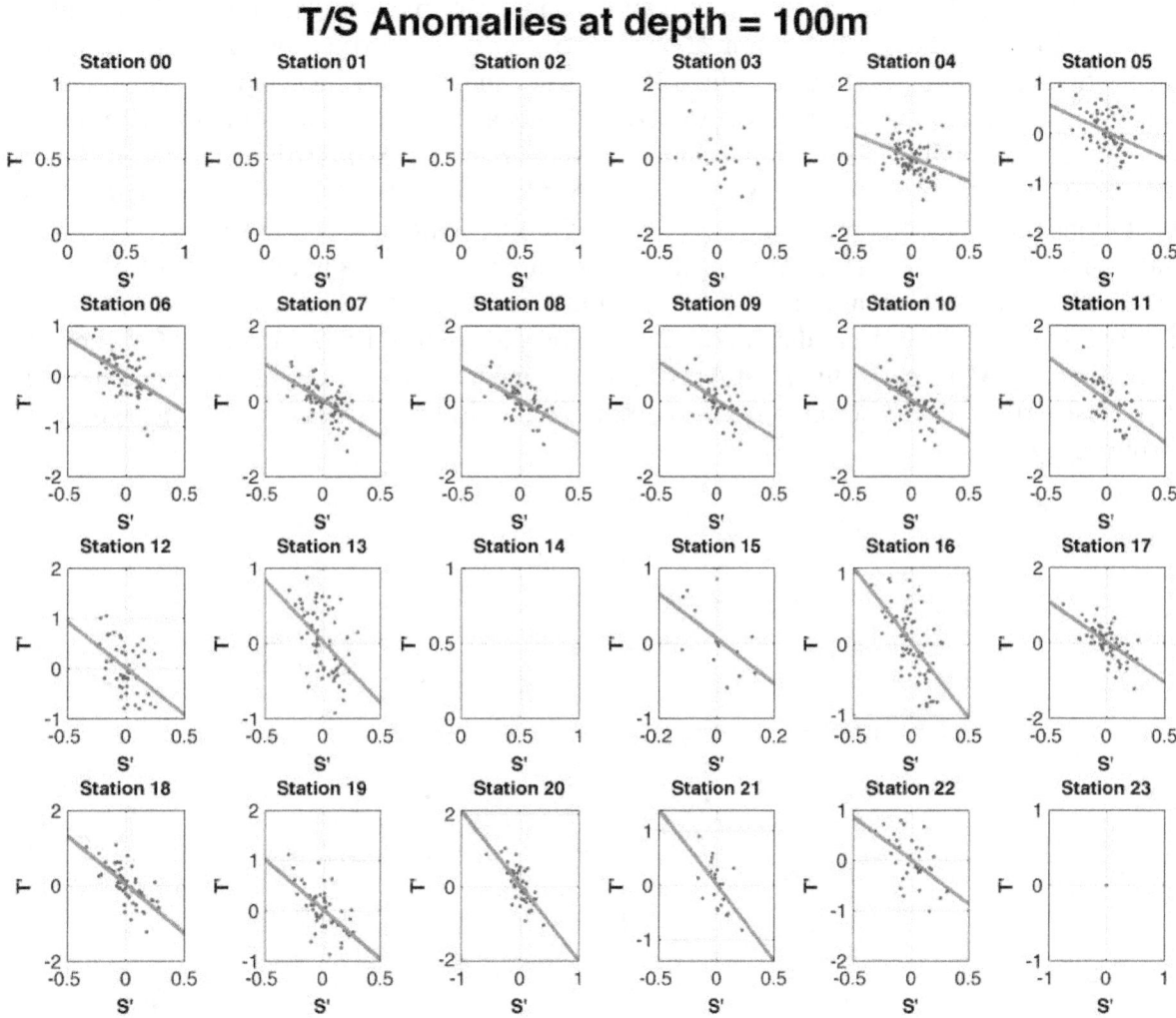

Figure 21. Anomalies of temperature (T') vs. anomalies of salinity (S') at 100 m depth for all stations. Significant relations at the 95% level are shown with a red best-fit linear trend line. Note that each panel is individually scaled to fit the range of observed anomalies.

From inspection of the anomaly time series shown above and those in appendices A, B and C it is clear that many anomaly signals extend across multiple stations and depth levels whereas others appear to be more tightly confined. In order to assess the strength and consistency of the relation between different depths, different stations, and different domains, we generate correlation tables at a number of discrete depth levels. We begin by examining the fluctuations of temperature and salinity within the water column at Stations 01, 04, 12 and 20. In the vertical direction, we find that temperature fluctuations tend to be coherent across the water column with the exception of 0 m depth variability, which is only weakly felt at depth at some stations and not felt at all at subsurface depths at Station 20 (Table 1). We note that 0 m depth measurements are difficult to make due to the presence of waves and air bubbles that can contaminate both the salinity and temperature measurements and so must be regarded with caution. Salinity variability (Table 2) is less strongly coherent than the temperatures through the water column (the distances of significant relations decrease).

In the horizontal dimension, temperature and salinity variability are strongly coherent at 150 m depth (Tables 3 and 4): only Station 22 shows no significant correlation to other measurement fluctuations (note that Station 22 is no longer part of the monitoring program and so has a smaller number of samples). Correlation coefficients at 10 m depth are much weaker between adjoining stations than at greater depths but most station still exhibit significant correlations for the temperature field, even those situated at either end of the fjord. Salinity anomalies, however, at Stations 00–02 are generally not correlated with fluctuations at stations north of the Central Bay Stations 05 and 13. This suggests that temperature anomalies tend to be controlled regionally while salinity anomalies are more localized phenomena. Rarely do salinity anomalies at 10 m depths located more than about 20 km apart account for more than 50% of the variability of each other. At 10 m depth, West Arm stations appear to be more tightly coupled to each other than East Arm stations; perhaps these two fjords have different rates of exchange with the Central Bay.

Table 1. Correlation table of temperature anomalies with depth at Stations 01, 04, 12 and 20; "---" denotes no significant relation found; "..." denotes no data available for that depth level.

```
STN DEPTH   0    10   20   30   50   75   100  150  200  250
01    0    1.00
01   10   0.79  1.00
01   20   0.72  0.95 1.00
01   30   0.60  0.81 0.94 1.00
01   50   0.50  0.58 0.69 0.78 1.00
01   75    ...   ...  ...  ...  ...  ...
01  100    ...   ...  ...  ...  ...  ...  ...
01  150    ...   ...  ...  ...  ...  ...  ...  ...
01  200    ...   ...  ...  ...  ...  ...  ...  ...  ...
01  250    ...   ...  ...  ...  ...  ...  ...  ...  ...  ...
04    0   1.00
04   10   0.25 1.00
04   20   0.26 0.81 1.00
04   30   0.17 0.65 0.84 1.00
04   50   0.08 0.47 0.63 0.84 1.00
04   75   0.06 0.35 0.49 0.68 0.90 1.00
04  100   ---  0.27 0.41 0.61 0.83 0.91 1.00
04  150   ---  0.18 0.31 0.48 0.64 0.74 0.87 1.00
04  200   ---  0.16 0.30 0.43 0.54 0.63 0.77 0.89 1.00
04  250   ---  0.12 0.23 0.35 0.38 0.47 0.63 0.78 0.90 1.00
12    0   1.00
12   10   0.48 1.00
12   20   0.30 0.78 1.00
12   30   0.23 0.63 0.88 1.00
12   50   0.17 0.51 0.74 0.91 1.00
12   75   0.11 0.34 0.50 0.69 0.83 1.00
12  100   0.13 0.31 0.44 0.62 0.76 0.96 1.00
12  150   0.13 0.28 0.39 0.52 0.62 0.79 0.82 1.00
12  200   0.10 0.10 0.16 0.27 0.34 0.58 0.63 0.86 1.00
12  250   0.10 0.11 0.15 0.21 0.25 0.43 0.42 0.68 0.82 1.00
20    0   1.00
20   10   ---  1.00
20   20   ---  0.71 1.00
20   30   ---  0.65 0.96 1.00
20   50   ---  0.57 0.85 0.94 1.00
20   75   0.09 0.55 0.74 0.84 0.94 1.00
20  100   ---  0.51 0.60 0.68 0.76 0.88 1.00
20  150   ---  0.45 0.58 0.65 0.67 0.75 0.81 1.00
20  200   ...   ...  ...  ...  ...  ...  ...  ...  ...
20  250   ...   ...  ...  ...  ...  ...  ...  ...  ...  ...
```

Table 2. Correlation table of salinity anomalies with depth at Stations 01, 04, 12 and 20; "—" denotes no significant relation found; "..." denotes no data available for that depth level.

```
STN DEPTH   0    10   20   30   50   75   100  150  200  250
01    0   1.00
01   10   0.72 1.00
01   20   0.45 0.78 1.00
01   30   —    0.21 0.56 1.00
01   50   —    —    0.12 0.53 1.00
01   75   ...  ...  ...  ...  ...  ...
01  100   ...  ...  ...  ...  ...  ...  ...
01  150   ...  ...  ...  ...  ...  ...  ...  ...
01  200   ...  ...  ...  ...  ...  ...  ...  ...  ...
01  250   ...  ...  ...  ...  ...  ...  ...  ...  ...  ...
04    0   1.00
04   10   —    1.00
04   20   —    0.43 1.00
04   30   —    0.20 0.69 1.00
04   50   —    0.12 0.50 0.78 1.00
04   75   —    0.13 0.45 0.67 0.92 1.00
04  100   —    0.10 0.45 0.63 0.88 0.96 1.00
04  150   —    0.07 0.34 0.49 0.72 0.81 0.86 1.00
04  200   —    0.08 0.28 0.39 0.57 0.64 0.66 0.85 1.00
04  250   —    —    0.22 0.29 0.37 0.44 0.44 0.67 0.88 1.00
12    0   1.00
12   10   0.25 1.00
12   20   —    0.47 1.00
12   30   —    0.32 0.77 1.00
12   50   0.17 0.25 0.56 0.80 1.00
12   75   0.10 0.25 0.49 0.63 0.71 1.00
12  100   0.14 0.20 0.38 0.49 0.63 0.88 1.00
12  150   0.17 0.14 0.29 0.37 0.61 0.71 0.85 1.00
12  200   —    0.09 0.20 0.17 0.29 0.40 0.54 0.74 1.00
12  250   0.12 0.15 0.25 0.23 0.32 0.39 0.54 0.71 0.85 1.00
20    0   1.00
20   10   —    1.00
20   20   —    0.39 1.00
20   30   —    0.41 0.70 1.00
20   50   —    0.14 0.23 0.65 1.00
20   75   —    —    0.15 0.47 0.87 1.00
20  100   —    —    0.16 0.40 0.75 0.85 1.00
20  150   —    —    —    0.26 0.60 0.66 0.79 1.00
20  200   ...  ...  ...  ...  ...  ...  ...  ...  ...
20  250   ...  ...  ...  ...  ...  ...  ...  ...  ...  ...
```

Table 3. Correlation of temperature at 10 m and 150 m depth levels between all station pairs; "—" denotes no significant relation found; "…" denotes no data available for that depth level.

DEPTH STN	00	01	02	03	04	05	06	07	08	09	10	11	12	13	14	15	16	17	18	19	20	21	22	23
10 00	1.00																							
10 01	0.48	1.00																						
10 02	0.44	0.70	1.00																					
10 03	0.39	0.61	0.78	1.00																				
10 04	0.29	0.35	0.39	0.58	1.00																			
10 05	0.24	0.35	0.47	0.49	0.58	1.00																		
10 06	0.27	0.51	0.53	0.57	0.50	0.61	1.00																	
10 07	0.13	0.21	0.23	0.22	0.39	0.40	0.72	1.00																
10 08	0.27	0.27	0.31	0.36	0.42	0.40	0.61	0.60	1.00															
10 09	0.36	0.36	0.42	0.49	0.39	0.40	0.65	0.14	0.57	1.00														
10 10	0.17	0.19	0.24	0.24	0.20	0.23	0.30	0.40	0.67	0.68	1.00													
10 11	0.14	0.18	0.23	0.19	0.09	0.23	0.15	0.53	0.47	0.64	1.00													
10 12	0.10	0.12	0.15	0.12	—	0.11	0.26	0.16	0.32	0.43	0.49	0.72	1.00											
10 13	0.36	0.40	0.54	0.49	0.70	0.76	0.56	0.49	0.48	0.37	0.26	0.11	0.08	1.00										
10 14	0.37	0.43	0.53	0.56	0.68	0.59	0.54	0.32	0.44	0.35	0.19	—	—	0.76	1.00									
10 15	0.27	0.21	0.35	0.42	0.49	0.46	0.26	0.16	—	—	0.20	—	—	0.60	0.65	1.00								
10 16	0.28	0.31	0.42	0.46	0.62	0.67	0.55	0.41	0.46	0.27	—	0.12	0.81	0.79	0.68	1.00								
10 17	0.30	0.31	0.40	0.41	0.54	0.68	0.61	0.38	0.41	0.42	0.29	0.10	0.21	0.75	0.65	0.56	0.75	1.00						
10 18	0.30	0.34	0.41	0.41	0.36	0.45	0.52	0.23	0.32	0.45	0.29	0.16	0.18	0.49	0.48	0.21	0.56	0.71	1.00					
10 19	0.18	0.18	0.28	0.29	0.25	0.39	0.36	0.12	0.25	0.46	0.25	0.24	0.32	0.42	0.35	0.15	0.45	0.63	0.65	1.00				
10 20	0.22	0.30	0.42	0.41	0.26	0.39	0.55	0.24	0.33	0.45	0.29	0.32	0.41	0.51	0.49	0.33	0.59	0.54	0.76	1.00				
10 21	—	0.18	0.24	0.25	—	—	0.27	0.30	0.47	0.62	0.63	0.71	0.80	0.18	—	—	0.21	0.19	0.39	0.41	1.00			
10 22	0.26	0.51	0.45	0.42	0.64	0.80	0.66	0.60	0.63	0.31	0.26	0.29	0.62	0.60	—	0.71	0.68	0.63	0.40	0.43	0.31	1.00		
10 23	0.31	0.36	0.40	0.31	0.43	0.61	0.73	0.62	0.67	0.68	0.38	0.34	0.45	0.56	0.47	—	0.59	0.67	0.57	0.50	0.36	0.45	0.83	1.00

DEPTH STN	00	01	02	03	04	05	06	07	08	09	10	11	12	13	14	15	16	17	18	19	20	21	22	23
150 00	…																							
150 01	…	…																						
150 02	…	…	…																					
150 03	…	…	…	…																				
150 04	…	…	1.00																					
150 05	…	…	0.71	1.00																				
150 06	…	…	0.65	0.68	1.00																			
150 07	…	…	0.74	0.73	0.87	1.00																		
150 08	…	…	0.69	0.69	0.84	0.85	1.00																	
150 09	…	…	0.65	0.75	0.78	0.85	0.90	1.00																
150 10	…	…	0.60	0.66	0.73	0.73	0.83	0.87	1.00															
150 11	…	…	0.59	0.62	0.57	0.67	0.74	0.80	0.83	1.00														
150 12	…	…	0.45	0.45	0.42	0.56	0.55	0.64	0.66	0.75	1.00													
150 13	…	…	…																					
150 14	…	…	…																					
150 15	…	…	…															1.00							
150 16	…	…	0.77	0.72	0.68	0.74	0.76	0.74	0.78	0.70							0.90	1.00						
150 17	…	…	0.66	0.65	0.71	0.69	0.70	0.72	0.75	0.58							0.92	0.91	1.00					
150 18	…	…	0.68	0.68	0.65	0.65	0.70	0.70	0.73	0.65							0.81	0.81	0.92	1.00				
150 19	…	…	0.58	0.58	0.50	0.52	0.57	0.57	0.68	0.59							0.88	0.81	0.86	0.81	1.00			
150 20	…	…	0.68	0.58	0.66	0.70	0.68	0.63	0.65	0.55							0.57	0.50	0.57	0.51	0.53	1.00		
150 21	…	…	0.43	0.43	0.46	0.48	0.57	0.59	0.63	0.55							—	—	—	—	—	—	1.00	
150 22	…	…	—	—	0.96	0.96	0.90			0.78						0.75	—	—	—	—	—	—	—	1.00
150 23	…	…	0.55	0.48	0.95																			

29

Table 4. Correlation of salinity at 10 m and 150 m depth levels between all station pairs; "---" denotes no significant relation found; "..." denotes no data available for that depth level.

```
DEPTH STN  00    01    02    03    04    05    06    07    08    09    10    11    12    13    14    15    16    17    18    19    20    21    22    23
10 00     1.00
10 01     0.17  1.00
10 02     0.11  0.38  1.00
10 03     ---   0.15  0.30  1.00
10 04     0.23  0.22  0.28  0.33  1.00
10 05     0.10  ---   0.10  0.18  0.24  1.00
10 06     ---   0.13  ---   0.39  0.11  0.33  1.00
10 07     0.07  ---   0.36  0.08  0.14  0.60  1.00
10 08     ---   ---   ---   0.13  0.60  0.66  1.00
10 09     ---   ---   0.21  ---   0.10  0.68  0.66  0.88  1.00
10 10     ---   ---   0.15  ---   ---   0.52  0.57  0.91  0.88  1.00
10 11     ---   ---   0.27  ---   ---   0.60  0.69  0.85  0.90  0.92  1.00
10 12     ---   ---   0.20  ---   ---   0.51  0.67  0.79  0.88  0.86  0.92  1.00
10 13     0.14  0.15  0.09  0.20  0.20  0.26  0.37  0.38  0.21  0.15  0.14  0.23  0.19  1.00
10 14     ---   ---   0.13  ---   0.13  0.31  0.42  0.48  0.39  0.52  0.48  0.36  0.34  1.00
10 15     ---   ---   ---   ---   0.38  0.24  ---   0.20  0.16  0.15  ---   ---   0.40  1.00
10 16     ---   ---   0.17  ---   0.10  0.48  0.59  0.62  0.59  0.68  0.54  0.42  0.68  0.36  1.00
10 17     ---   ---   ---   ---   ---   0.22  0.37  0.34  0.35  0.30  0.35  0.47  0.40  0.24  0.61  1.00
10 18     ---   0.08  ---   ---   ---   0.13  0.30  0.44  0.36  0.30  0.23  0.27  0.08  0.16  0.14  0.29  0.39  1.00
10 19     ---   ---   0.09  ---   ---   0.25  0.31  0.35  0.41  0.45  0.45  0.39  0.27  0.58  0.26  0.69  0.62  0.22  1.00
10 20     ---   ---   0.32  ---   ---   0.35  0.39  0.37  0.43  0.53  0.55  0.41  0.20  0.62  ---   0.70  0.32  ---   0.74  1.00
10 21     ---   ---   0.30  ---   ---   0.66  0.53  0.76  0.76  0.81  0.82  0.77  ---   0.21  ---   0.41  ---   ---   0.22  0.34  1.00
10 22     ---   0.13  ---   ---   ---   ---   ---   ---   ---   ---   ---   ---   ---   ---   ---   0.12  ---   0.22  0.19  ---   0.49  1.00
10 23     ---   ---   0.13  0.45  0.18  ---   ---   ---   ---   ---   ---   ---   ---   ---   ---   ---   ---   ---   ---   ---   ---   ---   ---
```

```
DEPTH STN  00    01    02    03    04    05    06    07    08    09    10    11    12    13    14    15    16    17    18    19    20    21    22    23
150 00    ...
150 01    ...   ...
150 02    ...   ...   ...
150 03    ...   ...   ...   ...
150 04    ...   ...   ...   ...  1.00
150 05    ...   ...   ...   ...  0.81  1.00
150 06    ...   ...   ...   ...  0.85  0.86  1.00
150 07    ...   ...   ...   ...  0.86  0.79  0.92  1.00
150 08    ...   ...   ...   ...  0.82  0.84  0.92  0.92  1.00
150 09    ...   ...   ...   ...  0.75  0.80  0.85  0.88  0.93  1.00
150 10    ...   ...   ...   ...  0.76  0.80  0.89  0.90  0.92  0.93  1.00
150 11    ...   ...   ...   ...  0.70  0.83  0.83  0.81  0.86  0.88  0.93  1.00
150 12    ...   ...   ...   ...  0.40  0.59  0.56  0.56  0.60  0.68  0.75  0.75  1.00
150 13    ...   ...   ...   ...   ...   ...   ...   ...   ...   ...   ...   ...   ...  ...
150 14    ...   ...   ...   ...   ...   ...   ...   ...   ...   ...   ...   ...   ...  ...  ...
150 15    ...   ...   ...   ...   ...   ...   ...   ...   ...   ...   ...   ...   ...  ...  ...  1.00
150 16    ...   ...   ...   ...  0.61  0.62  0.69  0.66  0.67  0.69  0.73  0.72  0.54  ...  ...  0.85  1.00
150 17    ...   ...   ...   ...  0.56  0.62  0.73  0.63  0.68  0.70  0.71  0.74  0.56  ...  ...  0.91  0.87  1.00
150 18    ...   ...   ...   ...  0.53  0.59  0.66  0.62  0.67  0.67  0.74  0.76  0.66  ...  ...  0.86  0.82  0.94  1.00
150 19    ...   ...   ...   ...  0.50  0.60  0.61  0.61  0.67  0.74  0.72  0.74  0.75  ...  ...  0.47  0.31  0.48  0.72  1.00
150 20    ...   ...   ...   ...  0.21  0.19  0.21  0.26  0.24  0.25  0.26  0.29  0.25  ...  ...  0.45  0.35  0.51  0.59  0.28  1.00
150 21    ...   ...   ...   ...  0.36  0.51  0.48  0.53  0.64  0.58  0.61  0.80  ...   ...  ...  ...   ...   ...   ...  0.87  1.00
150 22    ...   ...   ...   ...   ---  0.88   ---  0.80  0.96  0.74  0.93  0.94   ...  ...  ...  ...   ...   ...   ...   ...   ...   ...
150 23    ...   ...   ...   ...   ...   ...   ...   ...   ...   ...   ...   ...   ...  ...  ...  ...   ...   ...   ...   ...   ...   ...   ...
```

Relation to Other Observed Data and Climate Indices

In order to investigate possible causes of thermohaline anomalies in Glacier Bay waters, we regress monthly anomalies of the observed data against anomalies of the atmospheric variables shown in figure 7, the climate indices shown in figure 8 and observations of temperature and salinity recorded at oceanographic station GAK1. GAK1 is located at the mouth of Resurrection Bay, near Seward, Alaska, in 270 m water depth. It is the longest time series of coastal temperature and salinity in the coastal Alaska region and its data have been incorporated into numerous oceanographic studies, ecosystem assessments and fisheries forecast algorithms (see http://www.ims.uaf.edu/gak1).

We search for statistical significance between the Glacier Bay data and these various time series using linear regressions, examining the Pearson's R correlation coefficient along with the p-value appropriate for N-2 degrees of freedom where N is the number of unique monthly observations. This approach is somewhat sensitive to embedded autocorrelation within the data, so we caution that results may be sensitive to the presence of outliers.

For this analysis we select T and S observations from a near-surface depth level and a sub-surface depth level at each of our four focus stations (01, 04, 12 and 20). An example of the regression is shown in figure 22. We plot regression lines only for relations that exhibit a statistically significant relation at the 95% confidence level ($p<=0.05$). For the examples shown in figure 21, we note that warm GAK1 temperature anomalies near the bottom of the water column are coincident with cold temperature anomalies near the surface at Glacier Bay Station 20. In addition, the near-surface temperatures at this station are in phase with fluctuations of the PDO and are out of phase with the NPGO index. In all three of these cases, the fraction of total variance explained by the regression is weak ($<=12\%$) and only the NPGO relation is also significant at the 99% confidence level.

Other regression examples are shown in appendix E and a while few suggest some slight coupling between the parameters selected and Glacier Bay observations most show no relation. It is possible that lagged regressions incorporating some multi-month integrations may exhibit stronger relations but we do not pursue that avenue here. Multivariate regressions also can improve the fraction of variance accounted for beyond the predictive capability of these one-dimensional regressions. For example, when all environmental time series are considered together (selecting in order only those that explain a significant portion of the each residual time series variance) up to 35% of the total variance is accounted for at Station 12 at the 150 m depth level.

In aggregate, the impression that we get from these regressions is that Glacier Bay waters respond more strongly to local processes than the regional and remote processes represented by the time series chosen here. There are subtle indications that all of the selected climate indices do have some influence upon Glacier Bay's waters but the relations appear ephemeral (such as the warm signal associated with the 1997 El Niño) or weak overall. Additional work is needed in evaluate other processes not considered here, such as the role of ocean circulation response to wind forcing over the continental shelf and within the Southeast Alaska archipelago.

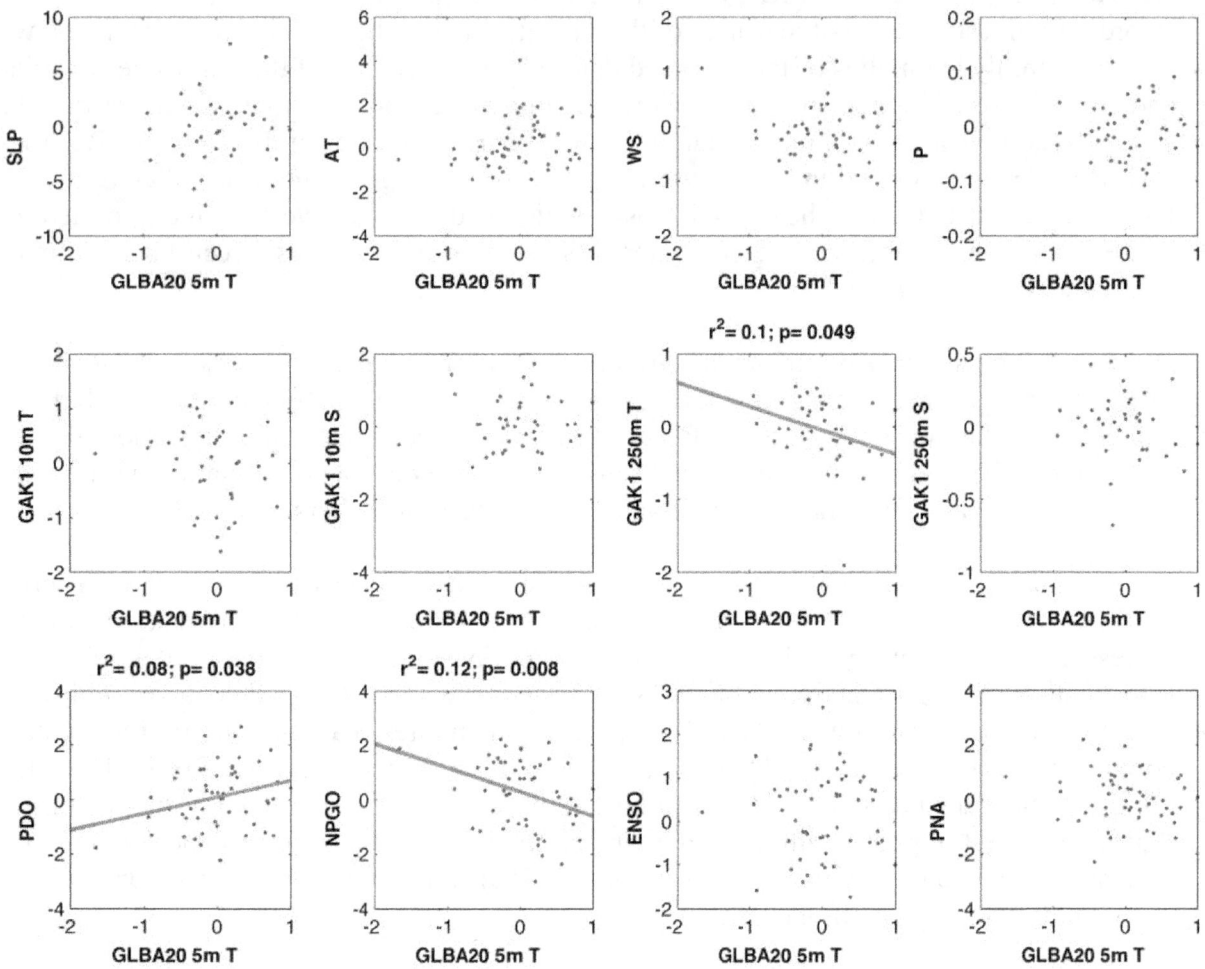

Figure 22. Scatter plots of Station 20 temperature anomalies at 5 m depth against 12 atmospheric, oceanographic and climate time series. Red lines denote linear regressions significant at the 95% confidence level or better; the statistical fit is described by the r^2 and p-values.

Discussion and Conclusions

This document provides a review of the measurements taken to date in the Glacier Bay oceanographic monitoring program. It is intended to provide an accessible manual to many processes that impact the regional northeast Pacific atmospheric and oceanographic conditions as well as provide a basic understanding of the character of the local variability. The appendices provide station-by-station depictions of the primary physical data on a seasonal basis and over the period of record.

Given the large apparent net production that takes places in Glacier Bay (inferred from the high density of marine mammals and birds that are supported within this relatively small region), the primary production of the bay must have some consistent source of nutrient supply. Because the renewal of deepwaters within the bay likely occurs primarily during the winter and spring months, much of this exchange takes place when there is not sufficient light to enable rapid phytoplankton growth. Figure 16 shows that waters outside of the bay above the sill depth are able to replace deepwaters between November and May but that intermediate waters could also exchange during summer months. Waters below the sill depth could likely replace deepwaters inside the fjord at any time of year. Certainly the nutrients supplying new production at the end of winter are dependent upon the resupply of deepwaters and winter wind mixing, however as this nutrient supply is drawn down over the course of the spring, where do the nutrients come from to replace these? The extent of intermediate water intrusion and the nutrient load of these waters should help at least bound this problem however the dynamics of nutrient resupply to the euphotic zone will require an identification of the location and rates of the mixing processes.

One possible mechanism for nutrient resupply is direct frictional drag and the interplay between frontal systems, stratification, advection and bottom friction. Another likely mechanism is that of internal wave breaking. In any estuary that hosts the combination of strong tidal currents and large variations in bathymetric depth, one would expect that internal wave generation would exert an appreciable control upon the resulting character of the density field. Internal wave generation likely occurs over each of the multiple sills found within Glacier Bay; it is less clear where the internal waves would propagate to and subsequently break. However, these sites of wave breaking could create natural hotspots for biological production. Whether nutrient resupply is due primarily to enhanced mixing due to frictional drag in shallow waters or whether it is due to internal wave dynamics is unknown.

It is clear that many issues remain to be resolved before a mechanistic understanding of Glacier Bay physical dynamics (and its mediation of the ecosystem) will be at hand. Some of these gaps in knowledge can be at least framed through application of theory and numerical modeling; most will require an observational effort that goes beyond the present ship-based monitoring program. The most direct route to a fuller understanding is through a set of instrumented moorings strategically stationed within and outside the fjord. Measurements of water velocity, temperature, salinity, light, light attenuation, fluorescence, and nutrients can be recorded with high frequency (~15–60 min.) over the course each deployment (typically one year) and so can capture the time evolution of the system at the mooring location in great detail.

We identify a number of relevant and pressing questions below. Some of these questions could be addressed with just one or two years of measurement; some would require multiple years of observation. Many are best addressed with moored instrumentation; others are better investigated with shipboard surveys or a combination of the two. A common thread linking all of these questions is the need to better understand what portions of the oceanographic variability are due to local dynamics and what portions of the variability are due to remote forcing (the first bullet item). All of these questions are also tied to the issue of residence times and deepwater renewal within the fjord (the second bullet item). By achieving this understanding, we will be in a better position to interpret the data collected by the ongoing oceanographic monitoring program.

- To what degree are oceanographic fluctuations in the greater Gulf of Alaska and within Icy Strait, Cross Sound and Chatham Strait felt within Glacier Bay?

- When does deepwater renewal occur? Continually in one portion of the year or in multiple pulses spread out in time?

- What sets the temperature, salinity and nutrient load of the replacement waters? Where do replacement waters come from?

- To what degree is biological production dependent upon the annual renewal of subsurface waters?

- Do the West Arm and the East Arm basins flush equally and in sync with one another?

- What leads to variability in the end-of-winter stratification profile? Do differences in stratification observed in February or March impact the subsequent spring bloom dynamics?

- How would Glacier Bay stratification look in the future with altered fresh water input from reduced glacial discharge? How might this affect the ecosystem?

- Where does nutrient exchange take place that supports primary production? Is it primarily fed by turbulent mixing from tides or are there regions of wind-driven upwelling that can spur mid-summer blooms? Is bottom friction or breaking internal waves more important?

- Is net production more dependent upon the resupply of nutrients to deep depths over the course of the winter or is it more dependent upon the resupply of nutrients to an intermediate layer on an ongoing basis through the summer?

Two types of mooring programs should be considered if the NPS decides that any of the above questions should be addressed: 1) incorporating a permanent fixed-location mooring into the monitoring program and 2) undertaking process-studies with mooring arrays deployed for 2–5 years in order to answer specific research questions. Moored observations from either type of program would add value to the shipboard CTD observations by placing the CTD measurements within a more complete temporal context and would add substantial value to both the measurements collected in the past and the measurements still to be taken in the future.

Literature Cited

Anderson, P. J., and J. F. Piatt. 1999. Community reorganization in the Gulf of Alaska following ocean climate regime shift. Marine Ecology Progress Series 189:117–123.

Boyer, T. P., J. I. Antonov, O. K. Baranova, H. E. Garcia, D. R. Johnson, R. A. Locarnini, A. V. Mishonov, T. D. O'Brien, D. Seidov, I. V. Smolyar, and others. 2009. World Ocean Database, 2009. In S. Levitus, editor. NOAA Atlas NESDIS 66. U.S. Government Printing Office, Washington, DC. 216 pp.

Danielson, S., W. Johnson, L. Sharman, G. Eckert, and B. Moynahan. 2010. Glacier Bay National Park and Preserve oceanographic monitoring protocol: Version OC-2010.1. Natural Resource Report NPS/SEAN/NRR—2010/265. National Park Service, Fort Collins, Colorado.

Di Lorenzo, E., N. Schneider, K. M. Cobb, K. Chhak, P. J. S. Franks, A. J. Miller, J. C. McWilliams, S. J. Bograd, H. Arango, E. Curchister, and others. 2008. North Pacific Gyre Oscillation links ocean climate and ecosystem change. Geophysical Research Letters 35, L08607, doi:10.1029/2007GL032838.

Etherington, L. 2006. Program evaluation report. U.S. Geological Survey, Alaska Science Center, Anchorage, Alaska.

Etherington, L. L., P. N. Hooge, E. R. Hooge, and D. F. Hill. 2007. Oceanography of Glacier Bay, Alaska: Implications for biological patterns in a glacial fjord estuary. Estuaries and Oceans 30(6):927–944.

Hill, D. F. 2007. Tidal modeling of Glacier Bay, Alaska - Methodology, results, and applications. Unpublished Report. Pennsylvania State University, University Park, Pennsylvania. 139 pp. Available from http://water.engr.psu.edu/hill/research/glba/default.stm (access 6 January 2012).

Hill, D. F., S. J. Ciavola, L. Etherington, and M. J. Klaar. 2009. Estimation of freshwater runoff into Glacier Bay, Alaska and incorporation into a tidal circulation model. Estuarine, Coastal and Shelf Science 82(1):95–107.

Hooge, P. N., and E. R. Hooge. 2002. Fjord oceanographic processes in Glacier Bay, Alaska. U.S. Geological Survey, Alaska Science Center, Anchorage, Alaska. 142 pp.

Hooge, P. N., E. R. Hooge, E. K. Solomon, C. L. Dezan, C. A. Dick, J. Mondragon, H. Rieden, and L. Etherington. 2003. Fjord oceanography monitoring handbook: Glacier Bay, AK. U.S. Geological Survey, Alaska Science Center, Anchorage, Alaska.

Janout, M. A., T. J. Weingartner, T. C. Royer, and S. L. Danielson. 2010. On the nature of winter cooling and the recent temperature shift on the northern Gulf of Alaska shelf. Journal of Geophysical Research 115, C05023, doi:10.1029/2009JC005774.

Mantua, N. J., S. R. Hare, Y. Zhang, J. M. Wallace, and R. C. Francis. 1997. A Pacific interdecadal climate oscillation with impacts on salmon production. Bulletin of the American Meteorological Society 78:1069–1079.

Matthews, J. B. 1981. The seasonal circulation of the Glacier Bay, Alaska fjord system. Estuarine, Coastal and Shelf Science 12(6):679–700.

Matthews, J. B., and A. V. Quinlan. 1975. Seasonal characteristics of water masses in Muir Inlet, a fjord with tidewater glaciers. Journal of the Fisheries Research Board of Canada 32:1693–1703.

Oakley, K. L., L. P. Thomas, and S. G. Fancy. 2003. Guidelines for long-term monitoring protocols. Wildlife Society Bulletin 31(4):1000–1003.

Pratt, L. 1986. Hydraulic control of sill flow with bottom friction. Journal of Physical Oceanography 16:1970–1980.

Royer, T. C. 2005. Hydrographic responses at a coastal site in the northern Gulf of Alaska to seasonal and interannual forcing, Deep Sea Research II 52(1–2):267–288, doi:10.1016/j.dsr2.2004.09.022.

Valle-Levinson, A., and R. E. Wilson. 1994. Effects of sill bathymetry, oscillating barotropic forcing and vertical mixing on estuary/ocean exchange. Journal of Geophysical Research 99(C3):5149–5169, doi:10.1029/93JC03333.

van den Dool, H. M., S. Saha, and Å Johansson. 2000. Empirical orthogonal teleconnections. Journal of Climate 13(8):1421–1435, doi: 10.1175/1520-0442(2000)013<1421:EOT>2.0.CO.

Weingartner, T. J., S. L. Danielson, and T. C. Royer. 2005. Freshwater variability and predictability in the Alaska Coastal Current. Deep Sea Research II 52:169–191, doi:10.1016/j.dsr2.2004.09.030.

Appendix A. Time-depth sections of temperature and salinity seasonal cycle and anomalies for the period of record over the upper 50 m.

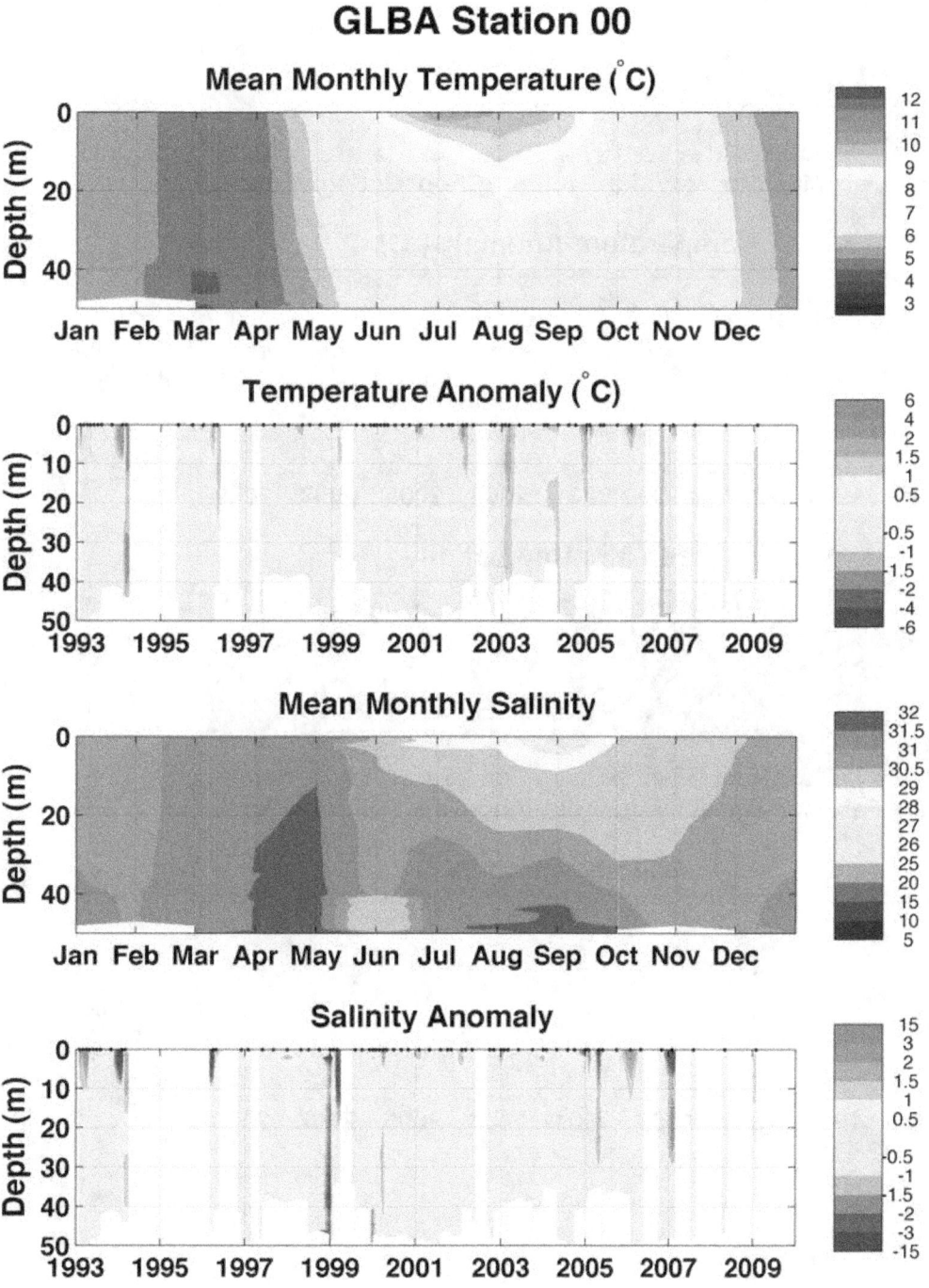

Glacier Bay Oceanographic Monitoring Program Analysis of Observations, 1993–2009
Appendix A. Time-depth sections of temperature and salinity seasonal cycle and anomalies for the period of record over the upper 50 m.

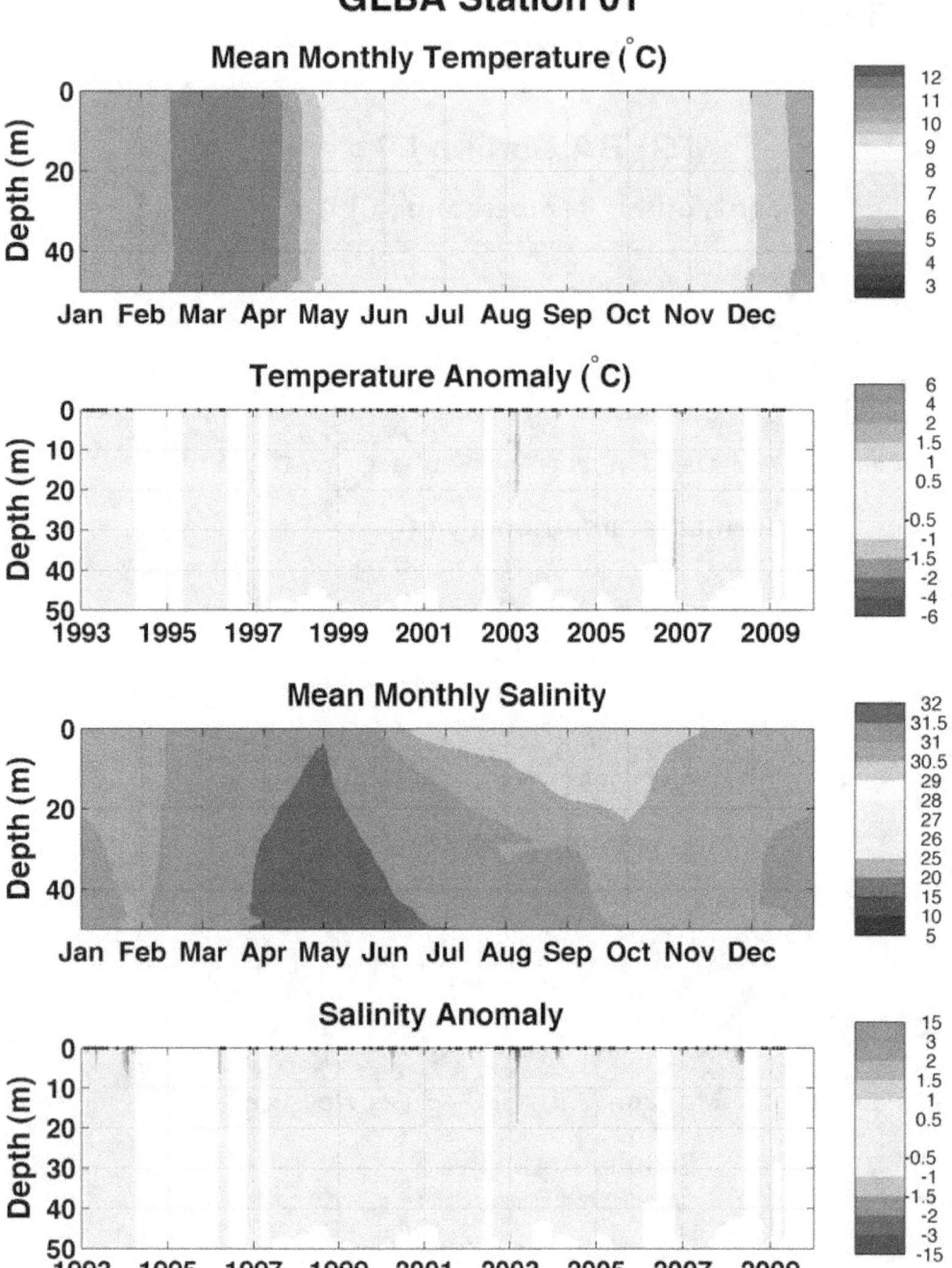

Glacier Bay Oceanographic Monitoring Program Analysis of Observations, 1993–2009
Appendix A. Time-depth sections of temperature and salinity seasonal cycle and anomalies for the period of record over the upper 50 m.

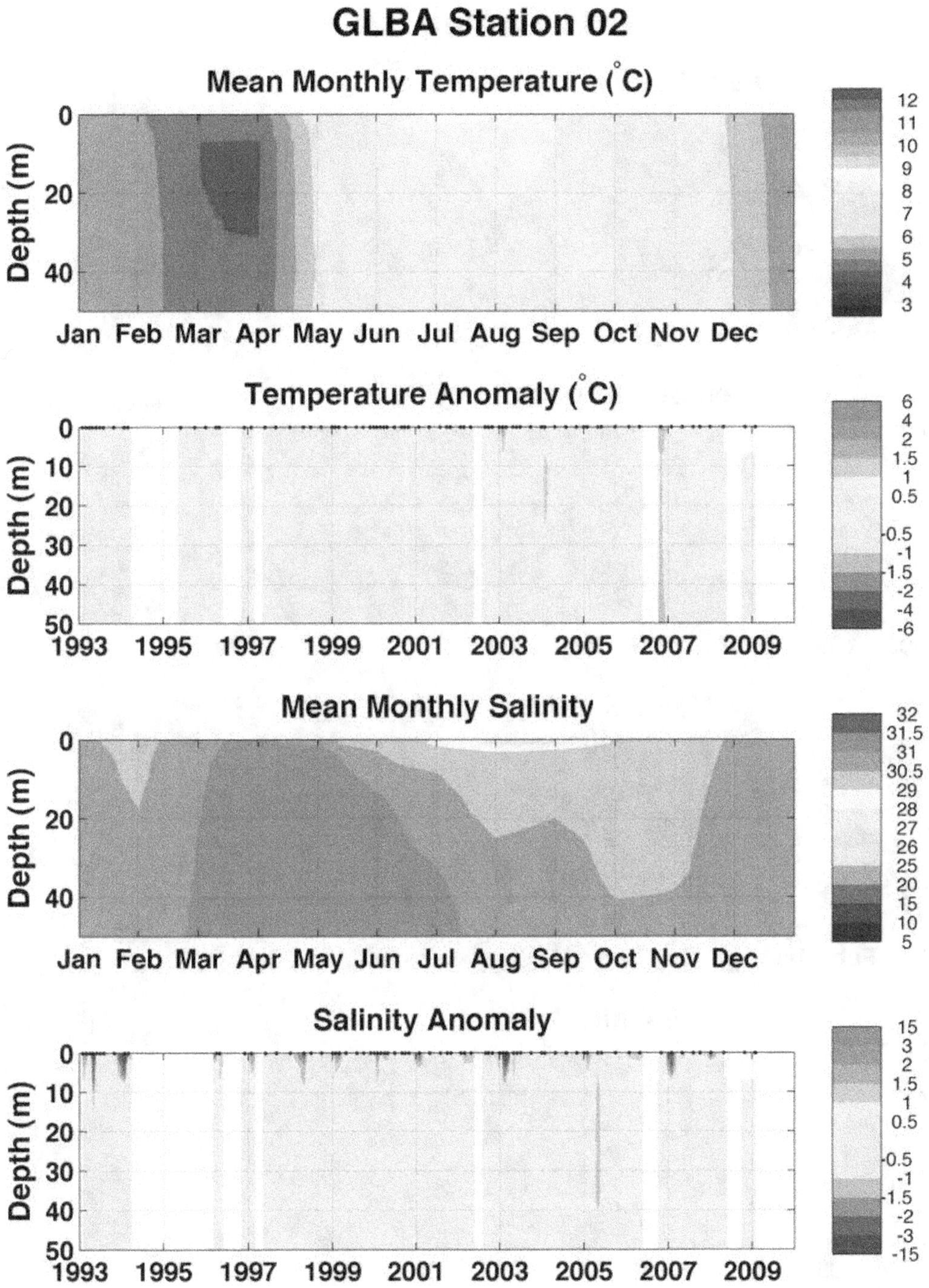

Glacier Bay Oceanographic Monitoring Program Analysis of Observations, 1993–2009
Appendix A. Time-depth sections of temperature and salinity seasonal cycle and anomalies for the period of record over the upper 50 m.

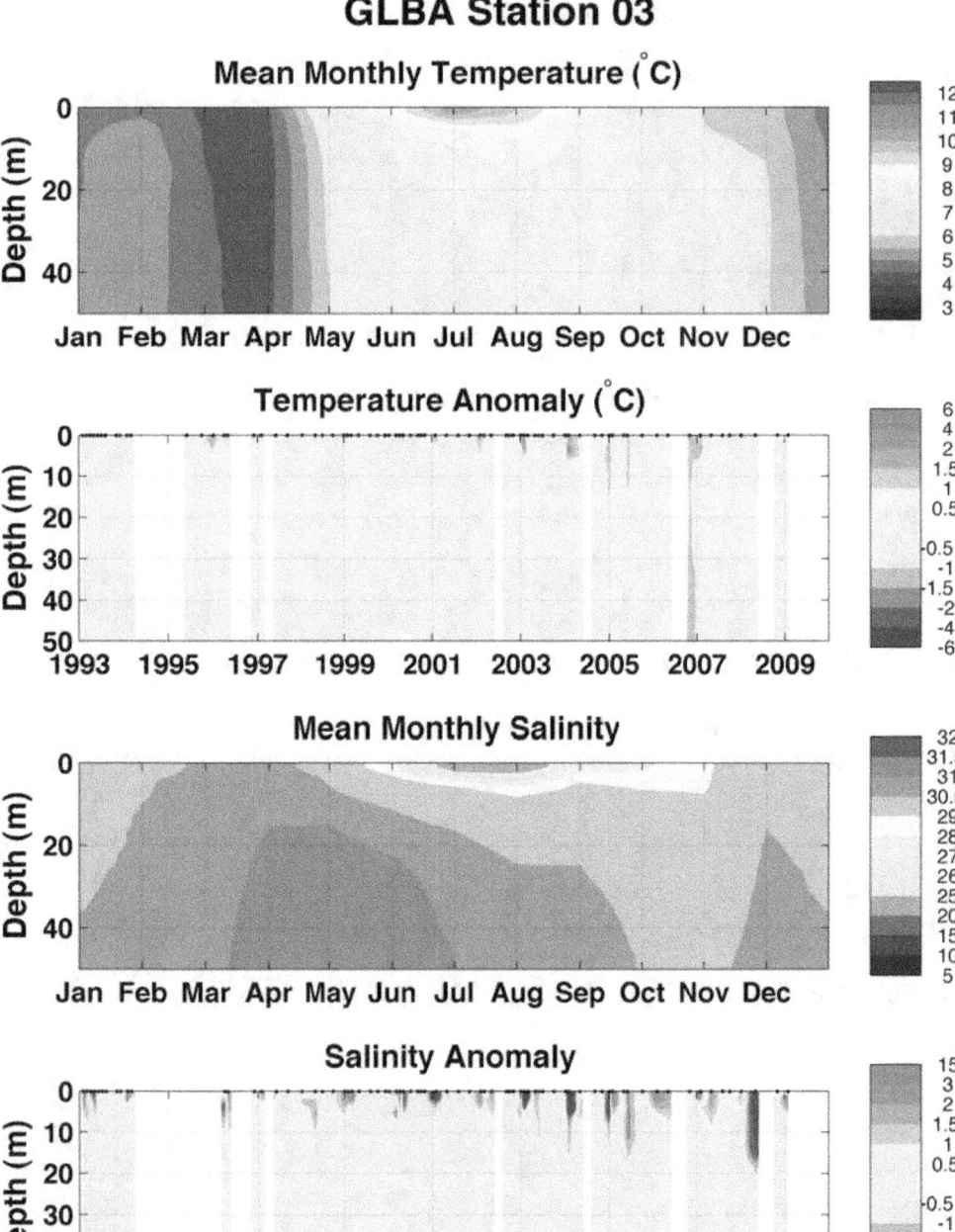

Glacier Bay Oceanographic Monitoring Program Analysis of Observations, 1993–2009
Appendix A. Time-depth sections of temperature and salinity seasonal cycle and anomalies for the period of record over the upper 50 m.

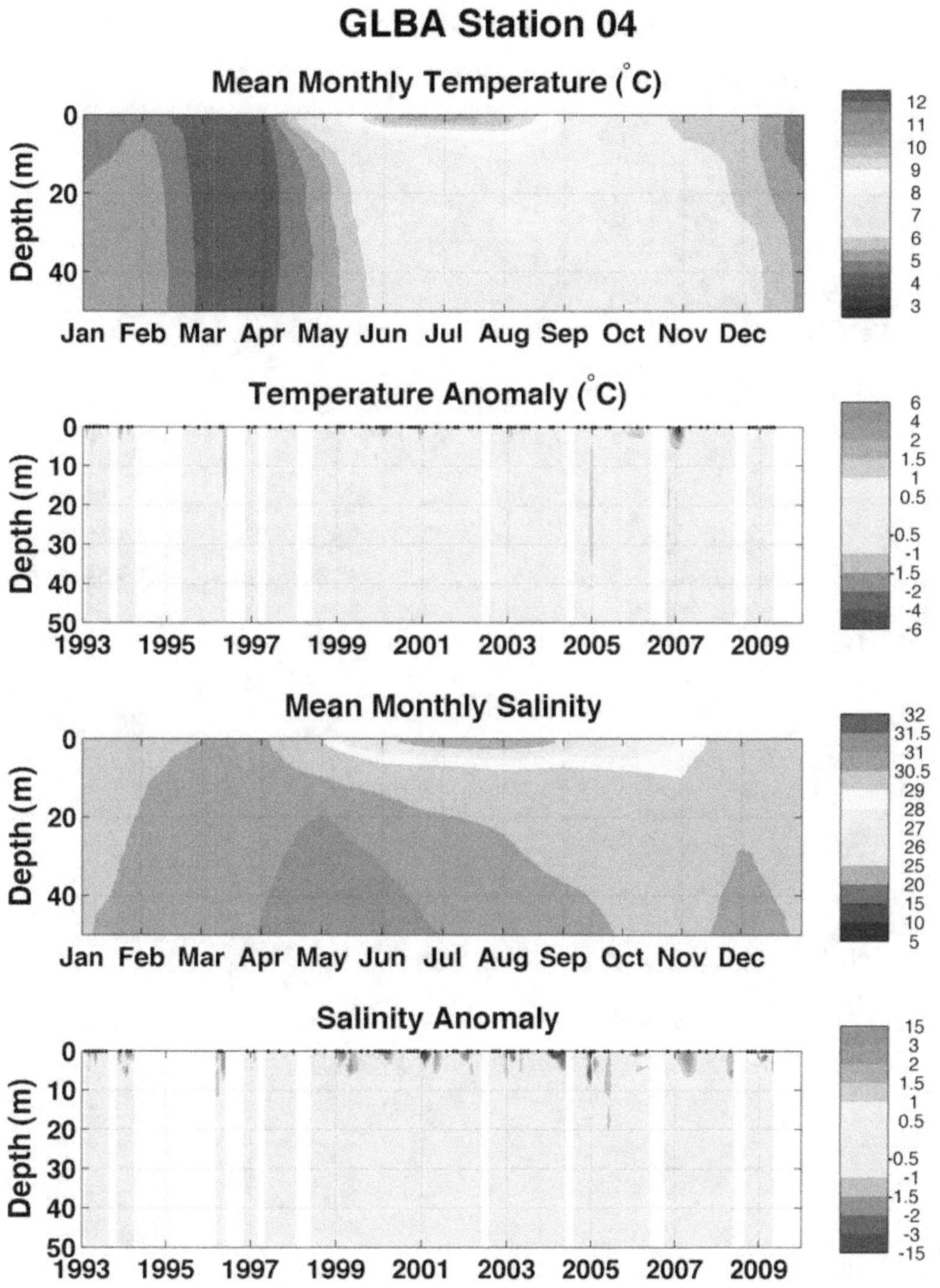

Glacier Bay Oceanographic Monitoring Program Analysis of Observations, 1993–2009
Appendix A. Time-depth sections of temperature and salinity seasonal cycle and anomalies for the period of record over the upper 50 m.

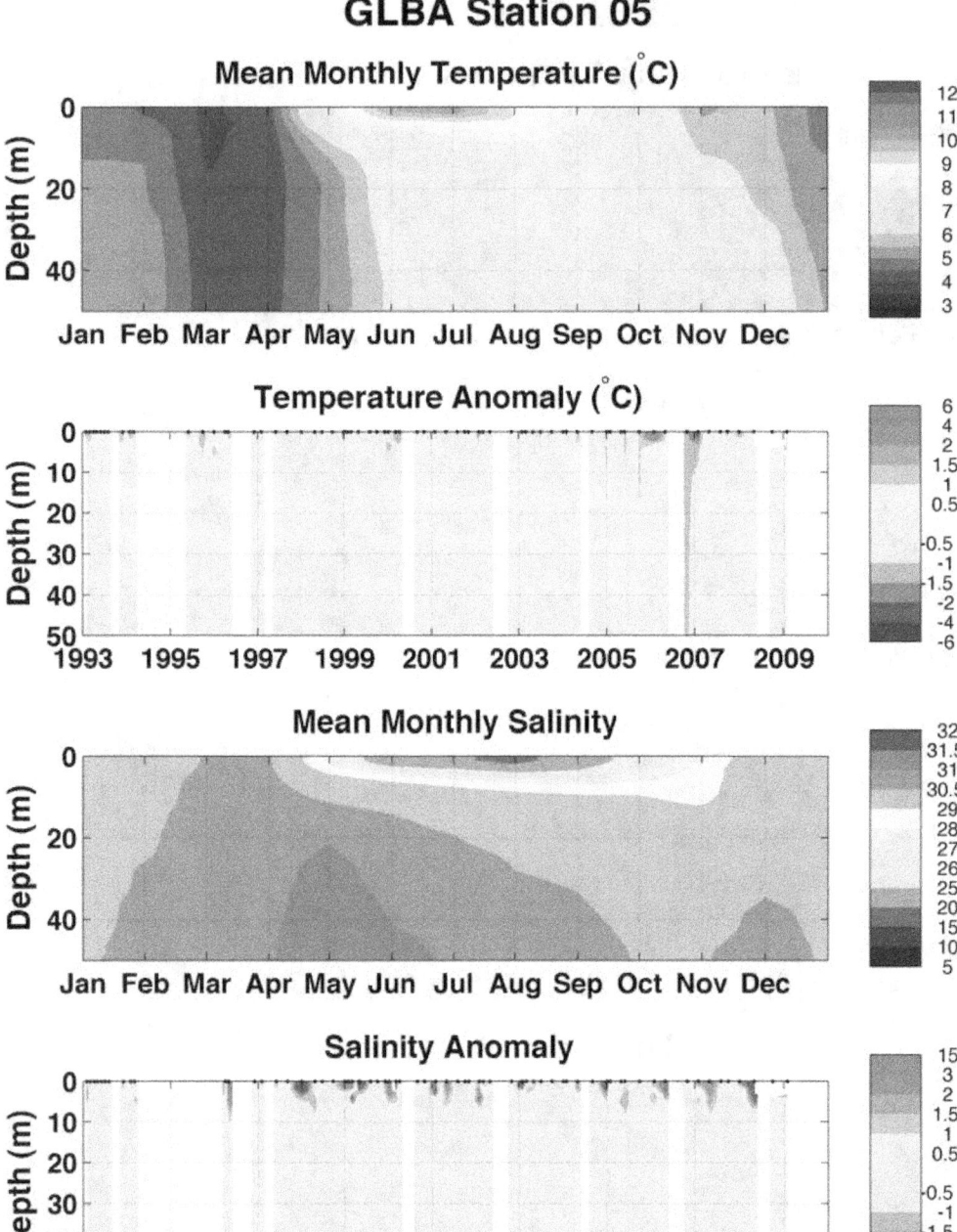

Glacier Bay Oceanographic Monitoring Program Analysis of Observations, 1993–2009
Appendix A. Time-depth sections of temperature and salinity seasonal cycle and anomalies for the period of record over the upper 50 m.

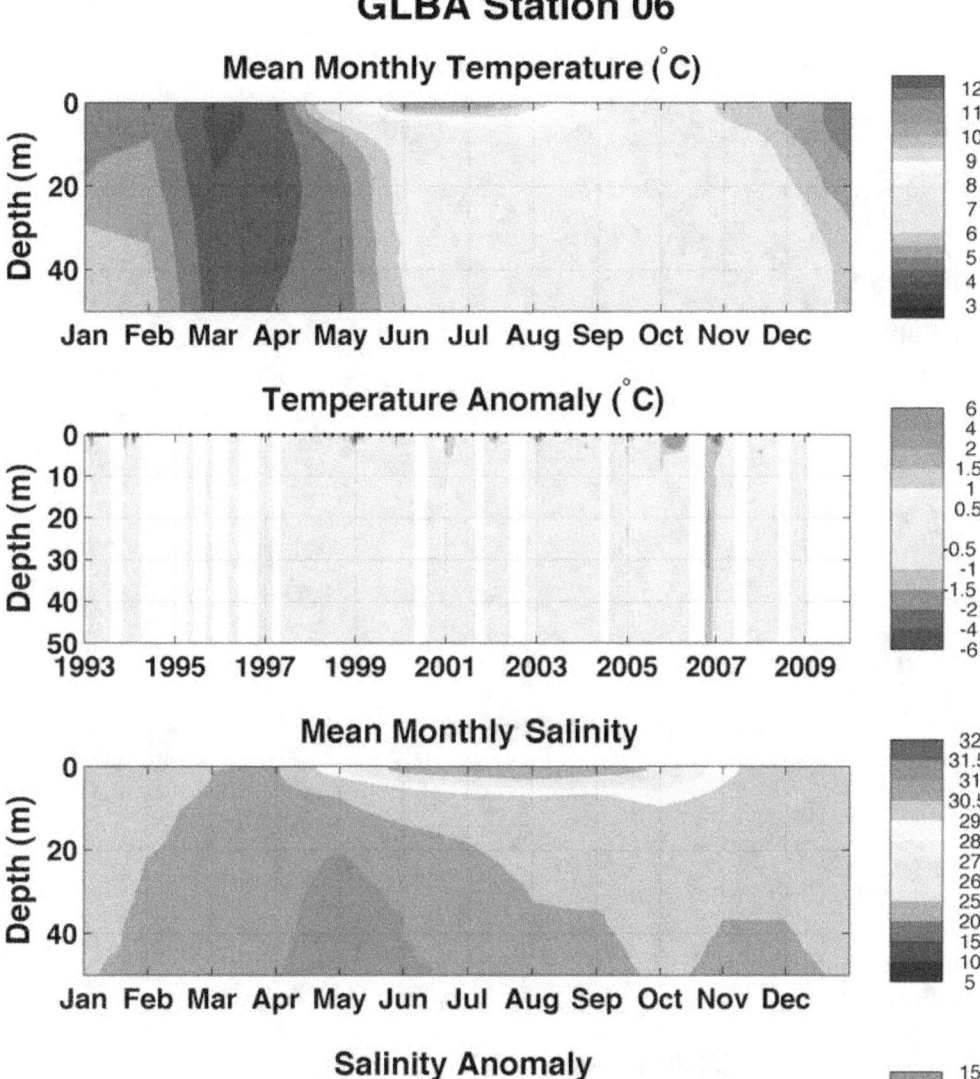

Glacier Bay Oceanographic Monitoring Program Analysis of Observations, 1993–2009
Appendix A. Time-depth sections of temperature and salinity seasonal cycle and anomalies for the period of record over the upper 50 m.

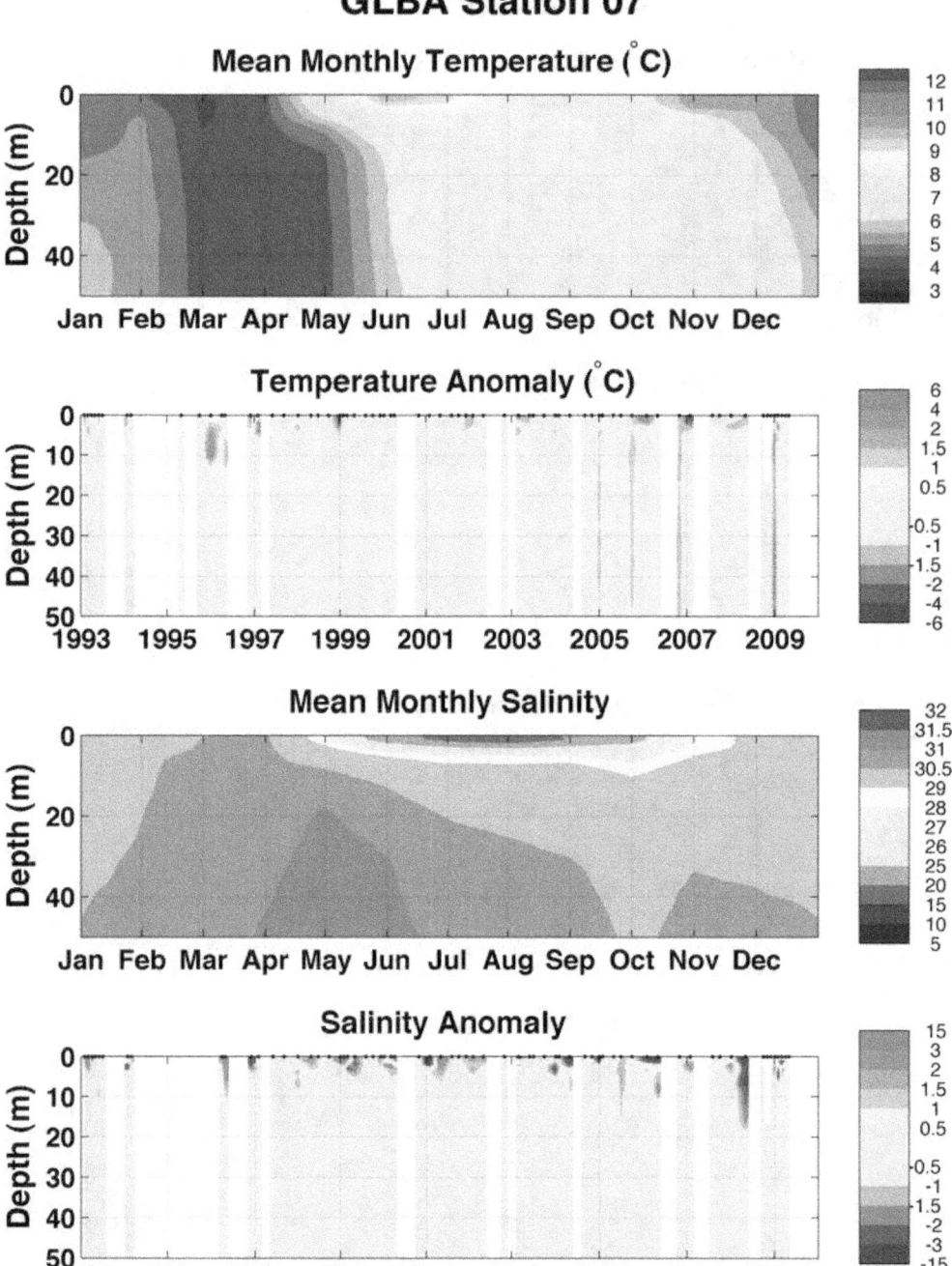

Glacier Bay Oceanographic Monitoring Program Analysis of Observations, 1993–2009
Appendix A. Time-depth sections of temperature and salinity seasonal cycle and anomalies for the period of record over the upper 50 m.

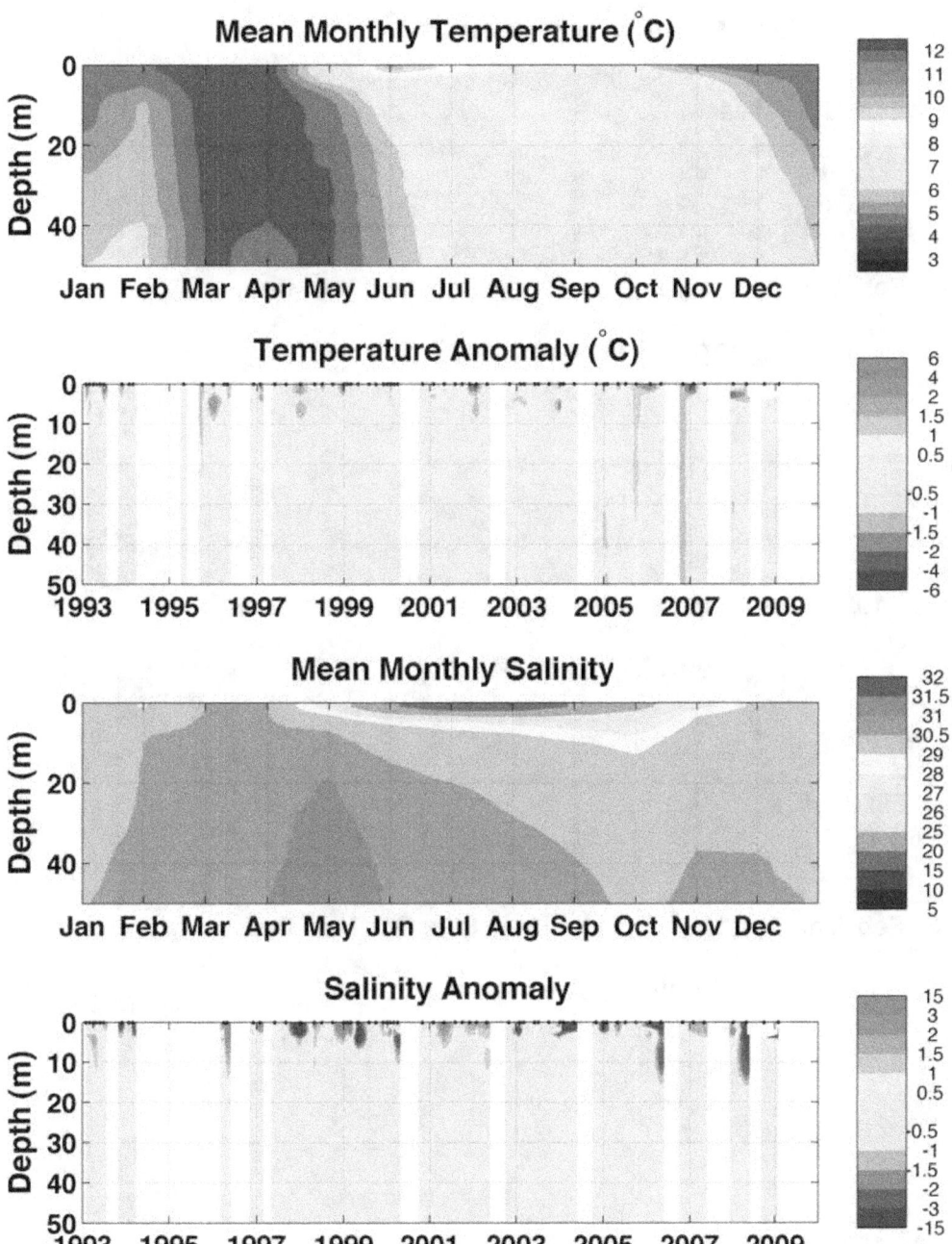

Glacier Bay Oceanographic Monitoring Program Analysis of Observations, 1993–2009
Appendix A. Time-depth sections of temperature and salinity seasonal cycle and anomalies for the period of record over the upper 50 m.

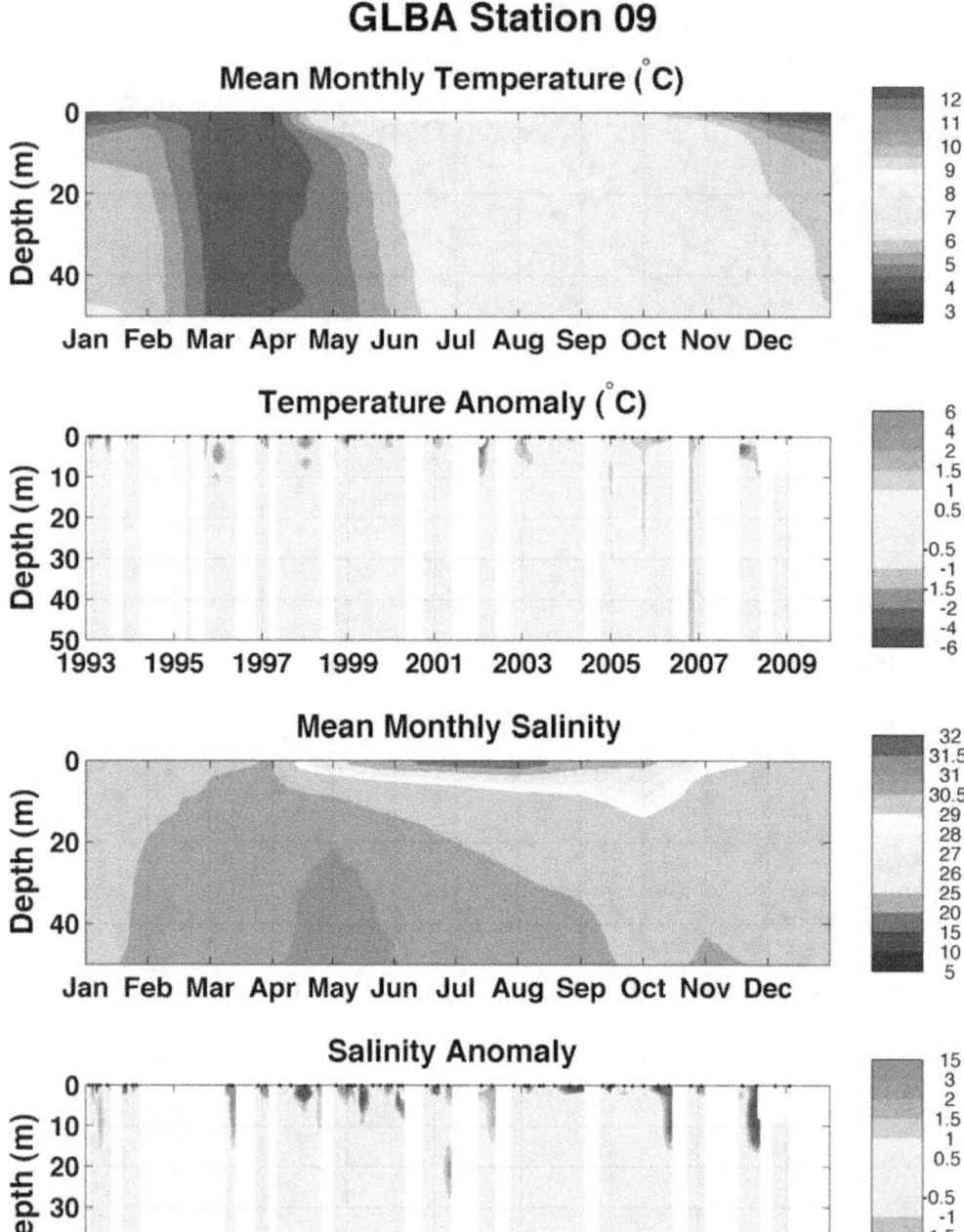

Glacier Bay Oceanographic Monitoring Program Analysis of Observations, 1993–2009
Appendix A. Time-depth sections of temperature and salinity seasonal cycle and anomalies for the period of record over the upper 50 m.

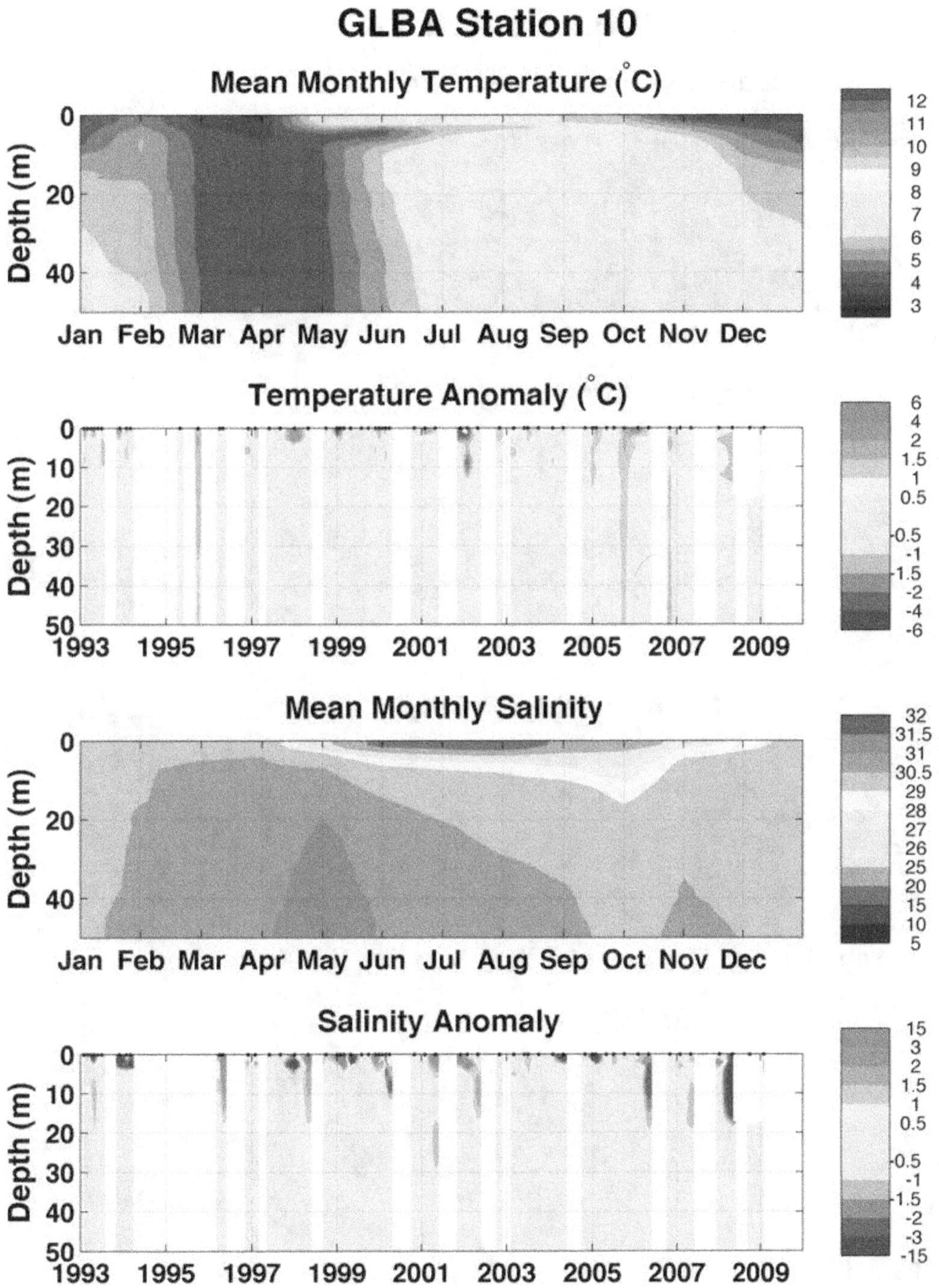

Glacier Bay Oceanographic Monitoring Program Analysis of Observations, 1993–2009
Appendix A. Time-depth sections of temperature and salinity seasonal cycle and anomalies for the period of record over the upper 50 m.

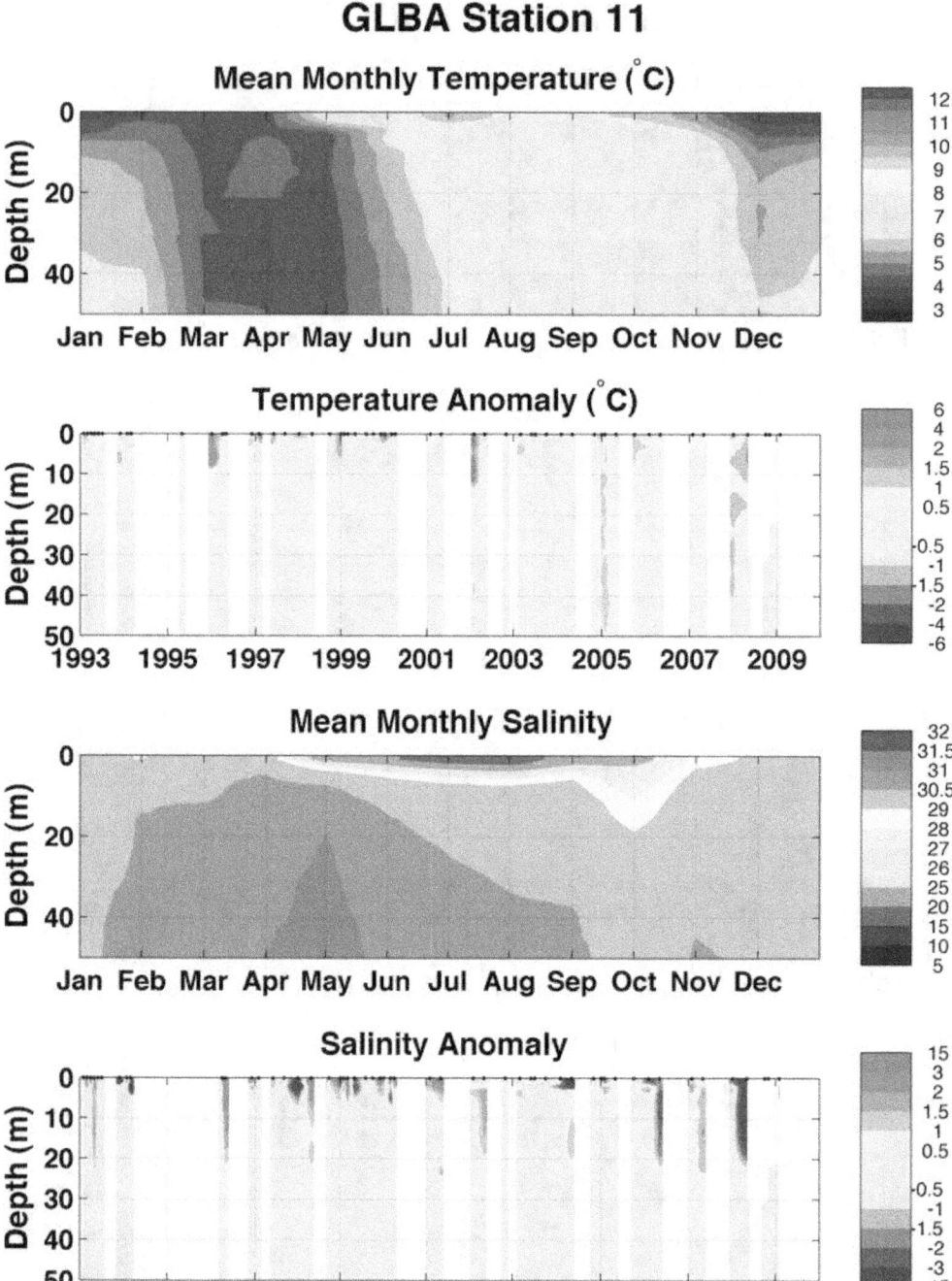

Glacier Bay Oceanographic Monitoring Program Analysis of Observations, 1993–2009
Appendix A. Time-depth sections of temperature and salinity seasonal cycle and anomalies for the period of record over the upper 50 m.

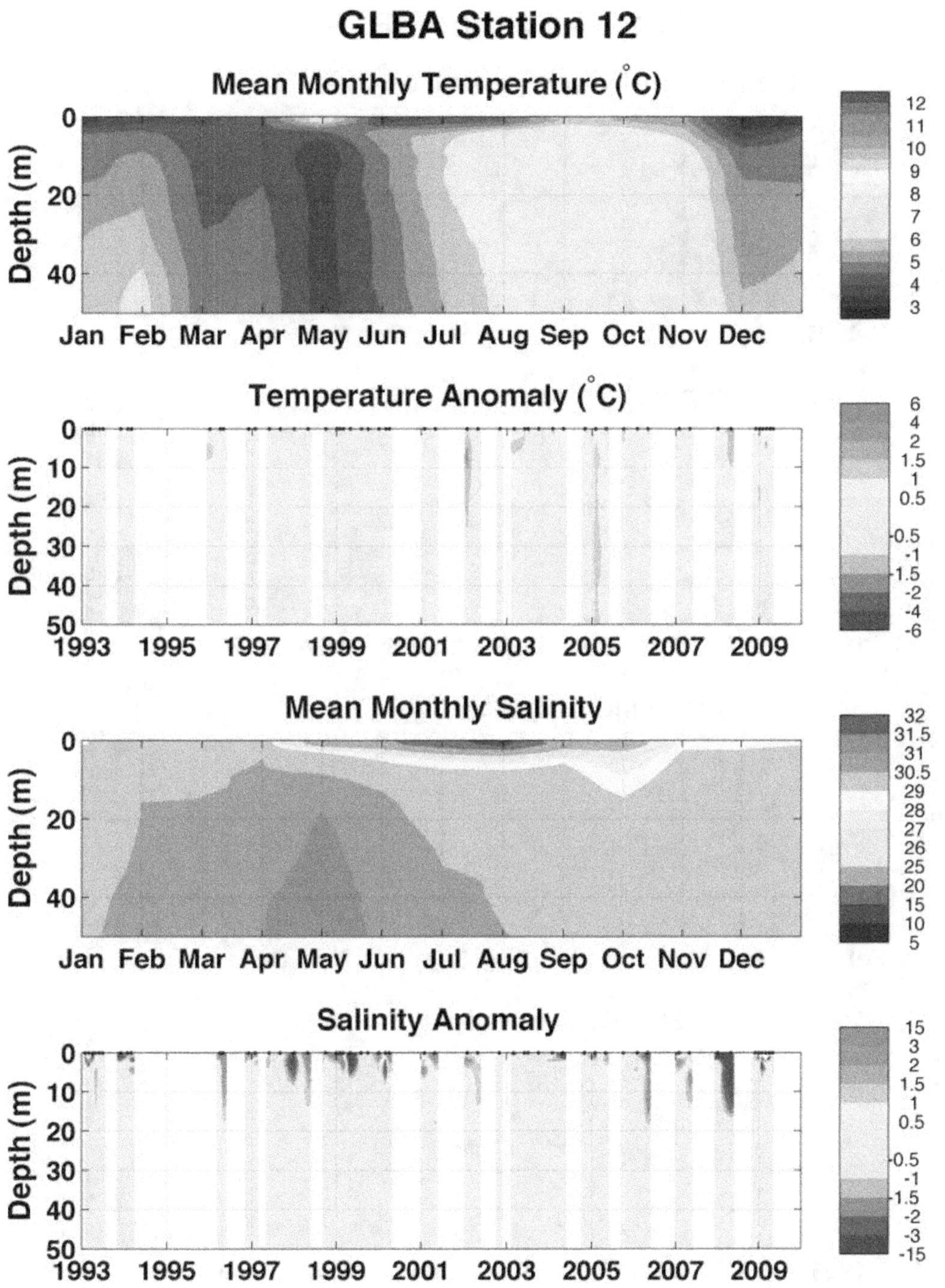

Glacier Bay Oceanographic Monitoring Program Analysis of Observations, 1993–2009
Appendix A. Time-depth sections of temperature and salinity seasonal cycle and anomalies for the period of record over the upper 50 m.

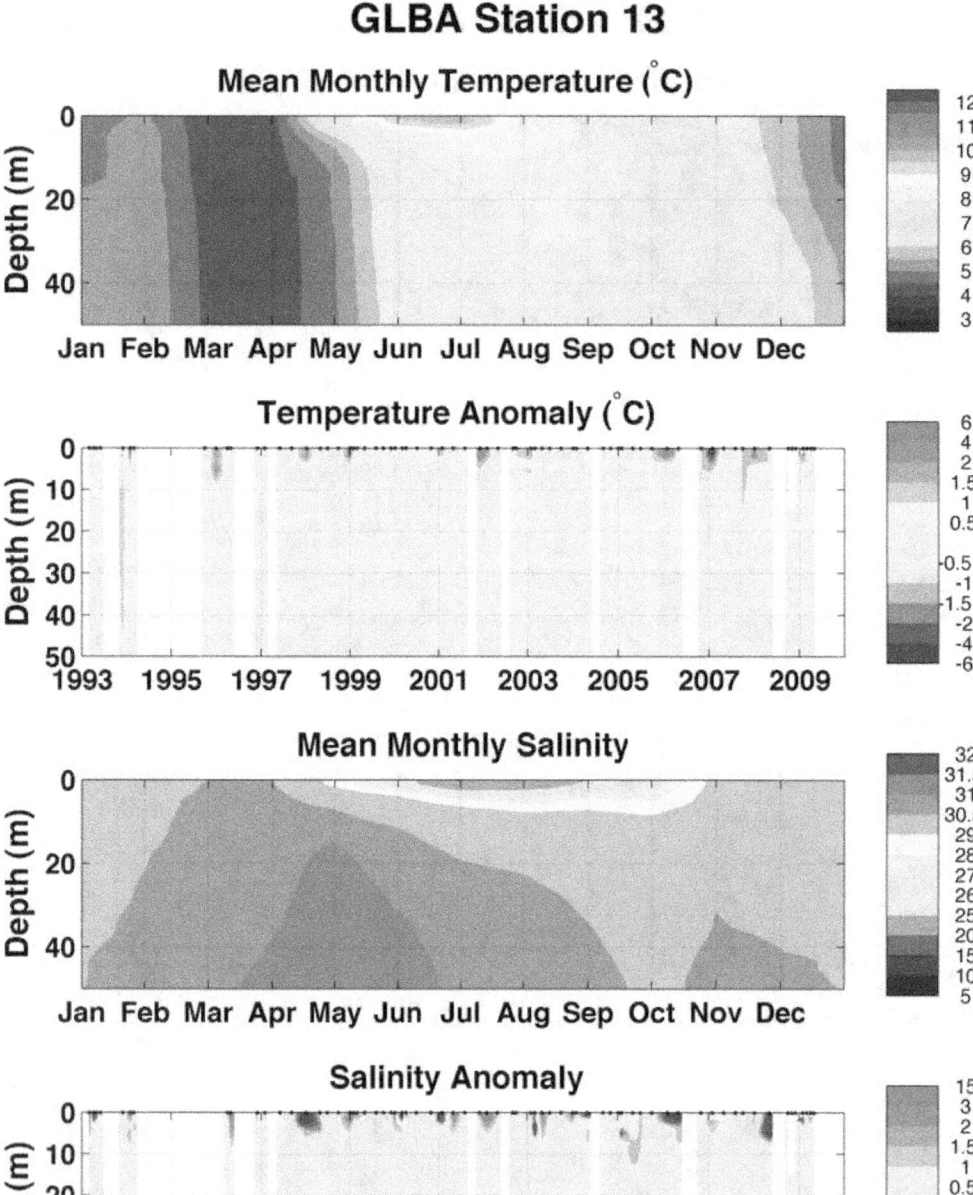

Glacier Bay Oceanographic Monitoring Program Analysis of Observations, 1993–2009
Appendix A. Time-depth sections of temperature and salinity seasonal cycle and anomalies for the period of record over the upper 50 m.

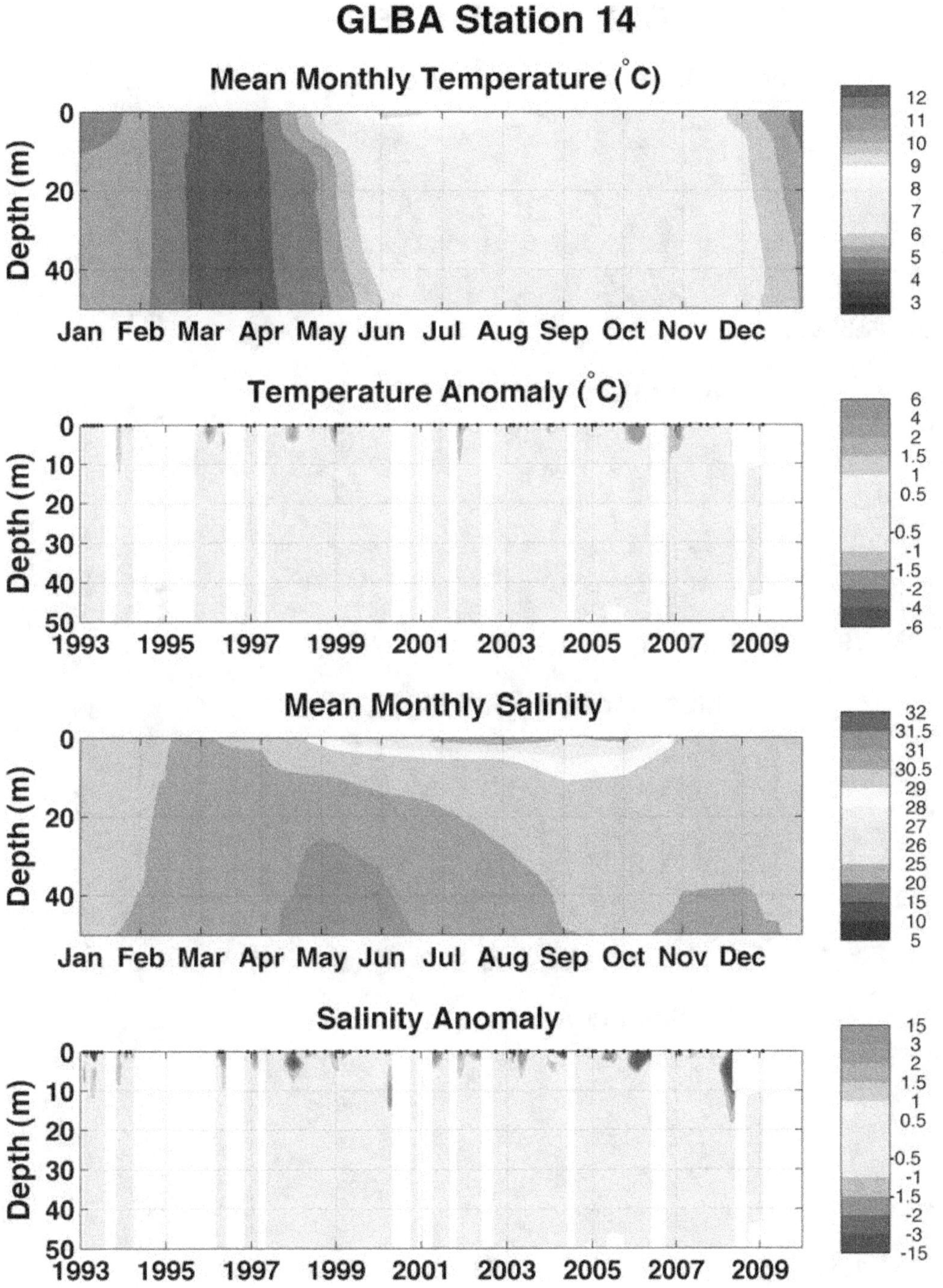

Glacier Bay Oceanographic Monitoring Program Analysis of Observations, 1993–2009
Appendix A. Time-depth sections of temperature and salinity seasonal cycle and anomalies for the period of record over the upper 50 m.

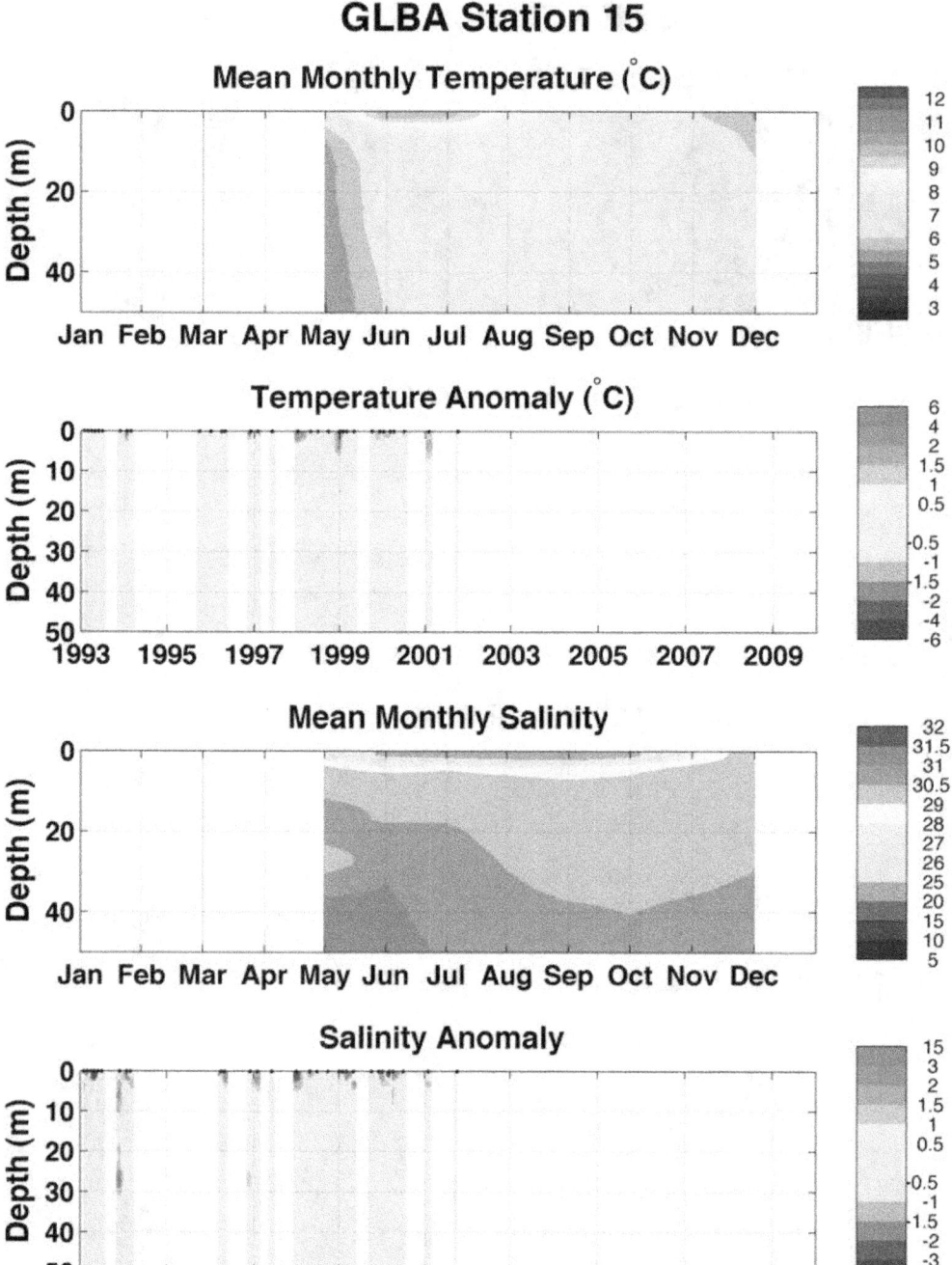

Glacier Bay Oceanographic Monitoring Program Analysis of Observations, 1993–2009
Appendix A. Time-depth sections of temperature and salinity seasonal cycle and anomalies for the period of record over the upper 50 m.

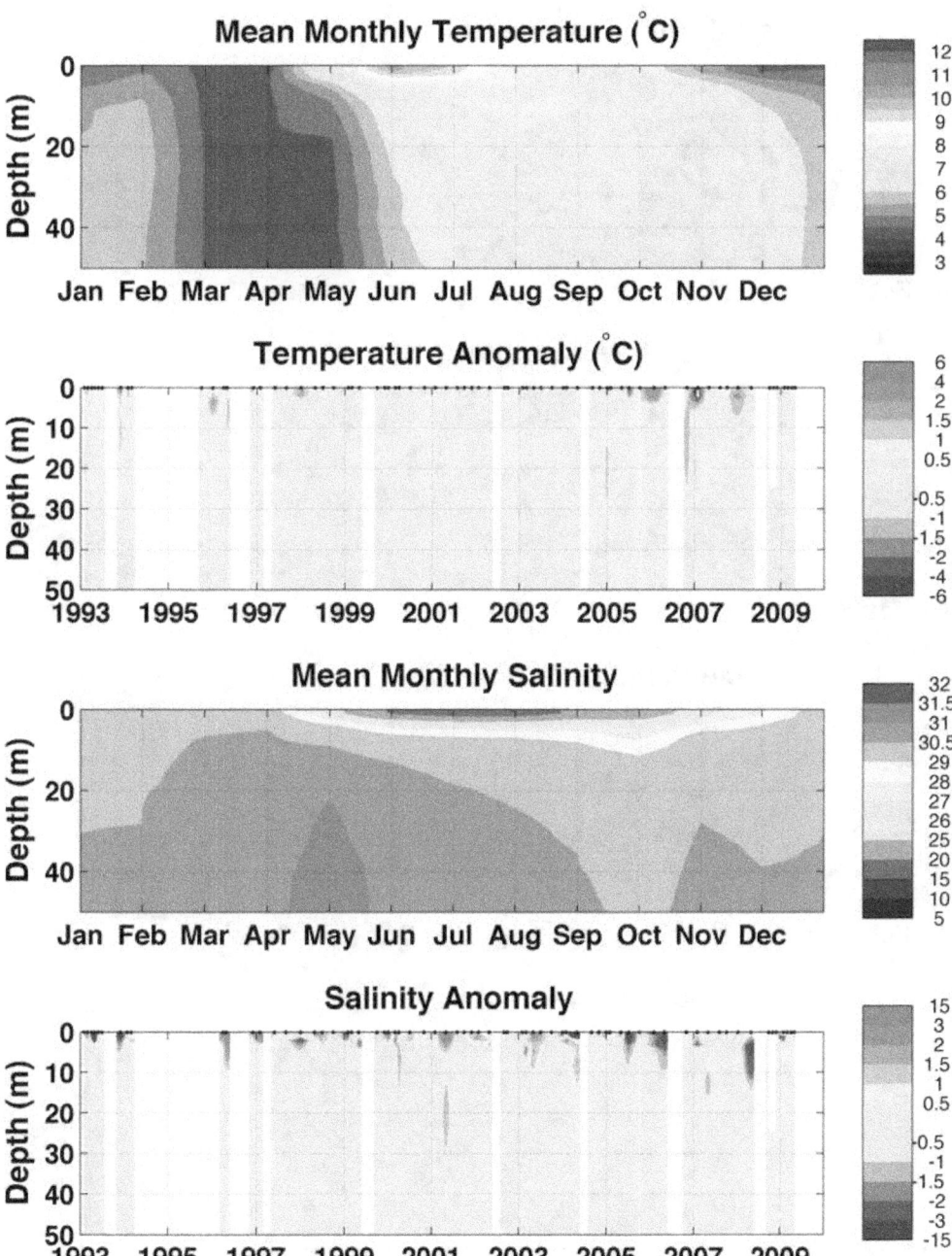

Glacier Bay Oceanographic Monitoring Program Analysis of Observations, 1993–2009
Appendix A. Time-depth sections of temperature and salinity seasonal cycle and anomalies for the period of record over the upper 50 m.

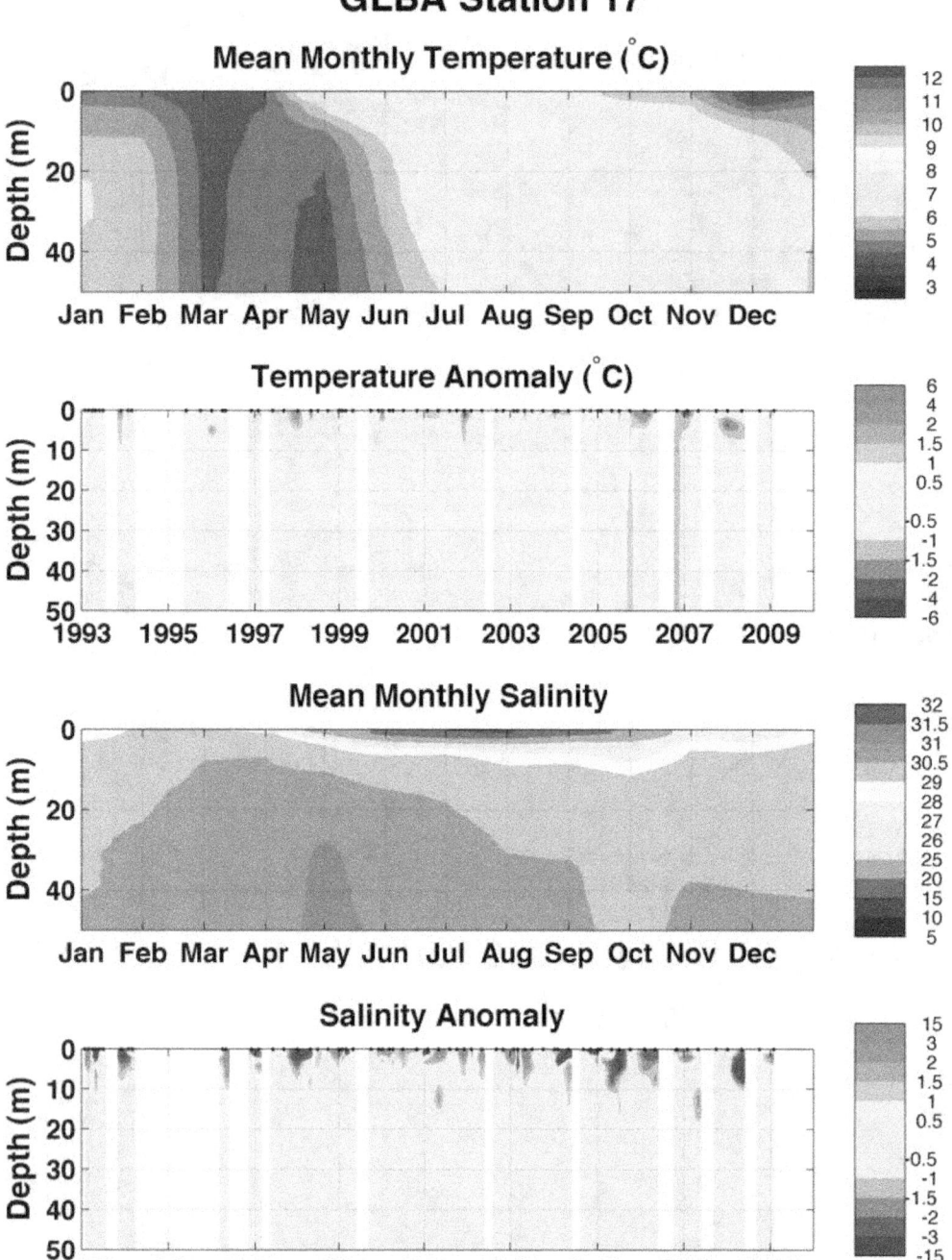

Glacier Bay Oceanographic Monitoring Program Analysis of Observations, 1993–2009
Appendix A. Time-depth sections of temperature and salinity seasonal cycle and anomalies for the period of record over the upper 50 m.

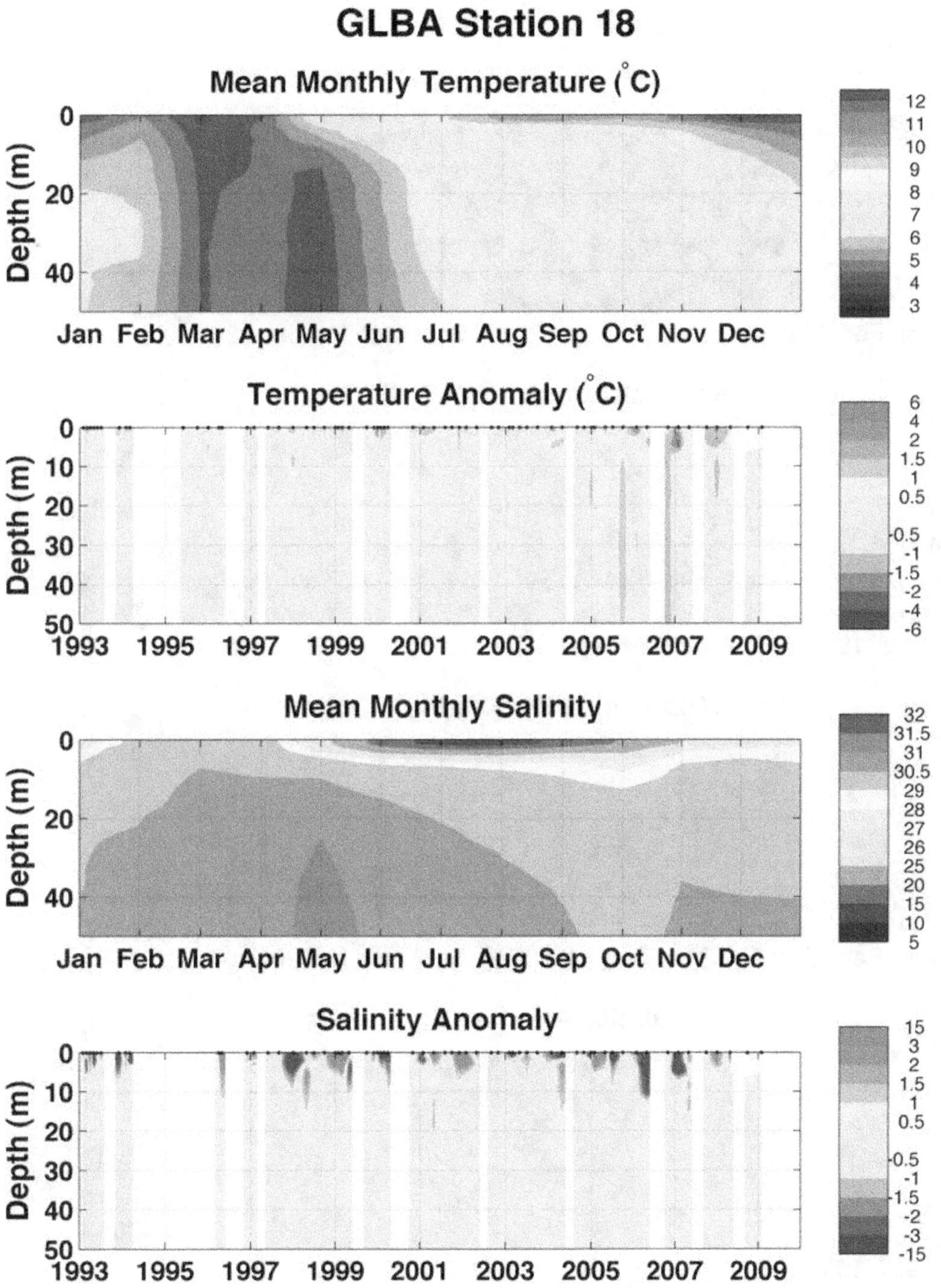

Glacier Bay Oceanographic Monitoring Program Analysis of Observations, 1993–2009
Appendix A. Time-depth sections of temperature and salinity seasonal cycle and anomalies for the period of record over the upper 50 m.

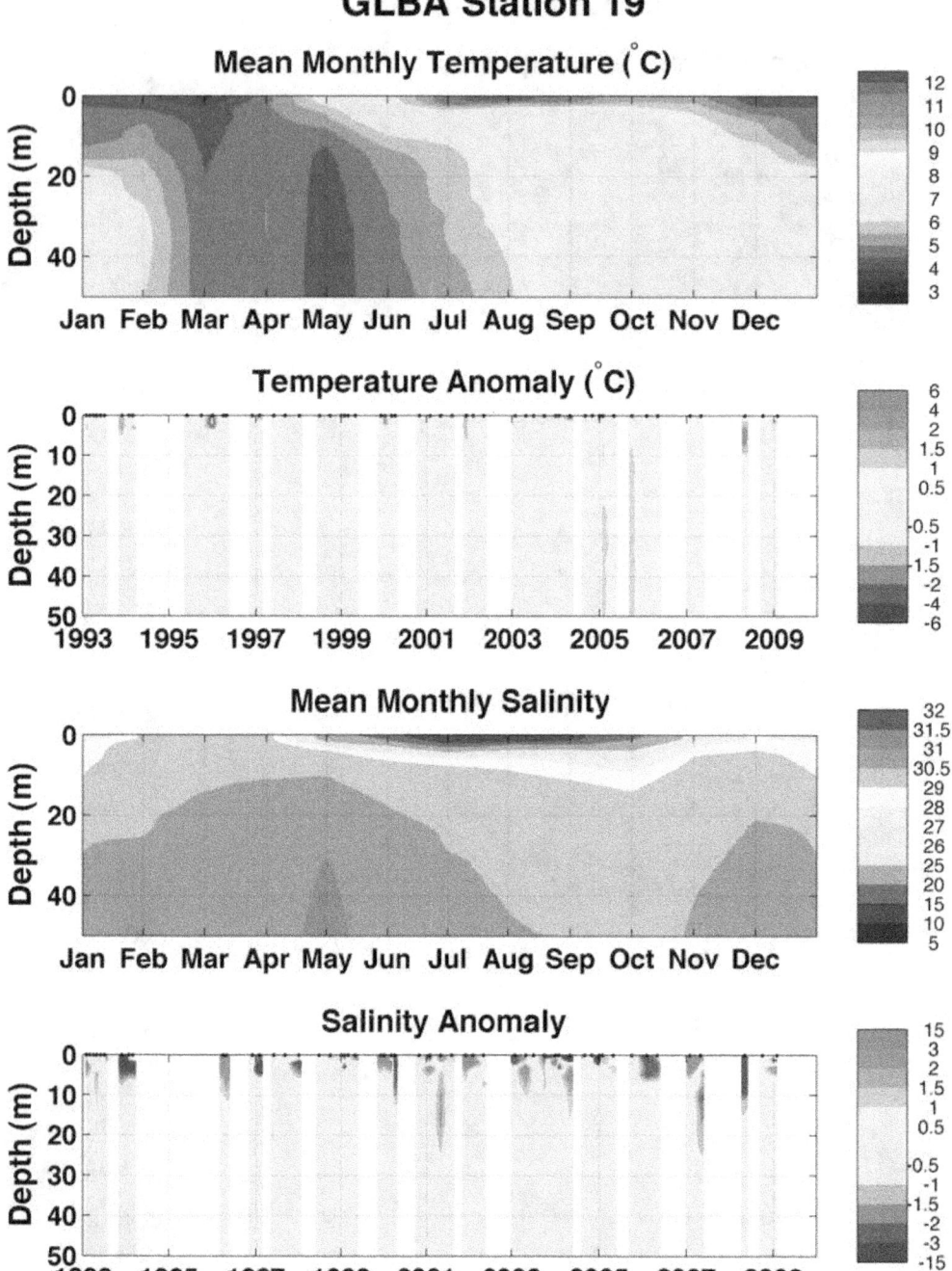

Glacier Bay Oceanographic Monitoring Program Analysis of Observations, 1993–2009
Appendix A. Time-depth sections of temperature and salinity seasonal cycle and anomalies for the period of record over the upper 50 m.

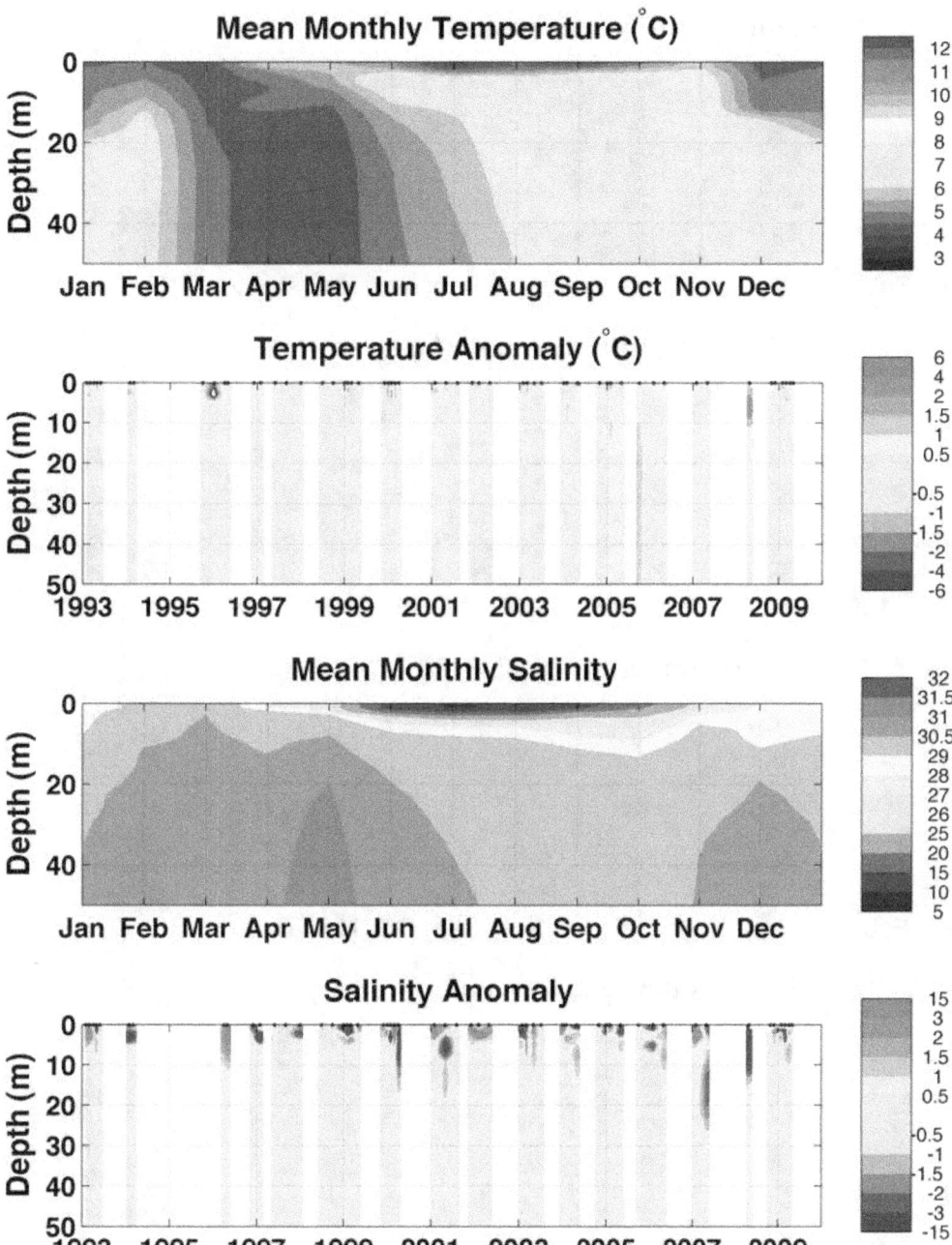

Glacier Bay Oceanographic Monitoring Program Analysis of Observations, 1993–2009
Appendix A. Time-depth sections of temperature and salinity seasonal cycle and anomalies for the period of record over the upper 50 m.

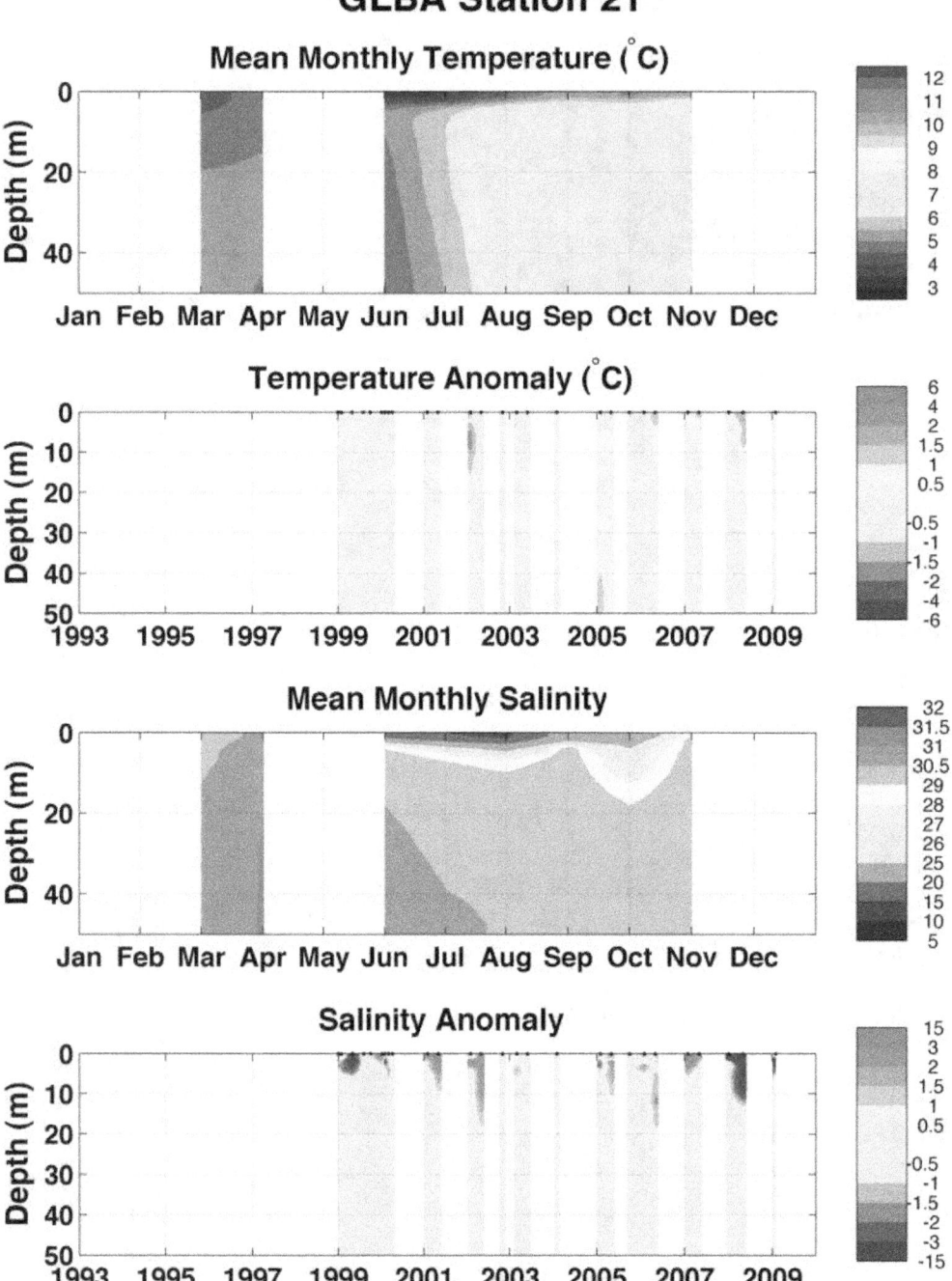

Glacier Bay Oceanographic Monitoring Program Analysis of Observations, 1993–2009
Appendix A. Time-depth sections of temperature and salinity seasonal cycle and anomalies for the period of record over the upper 50 m.

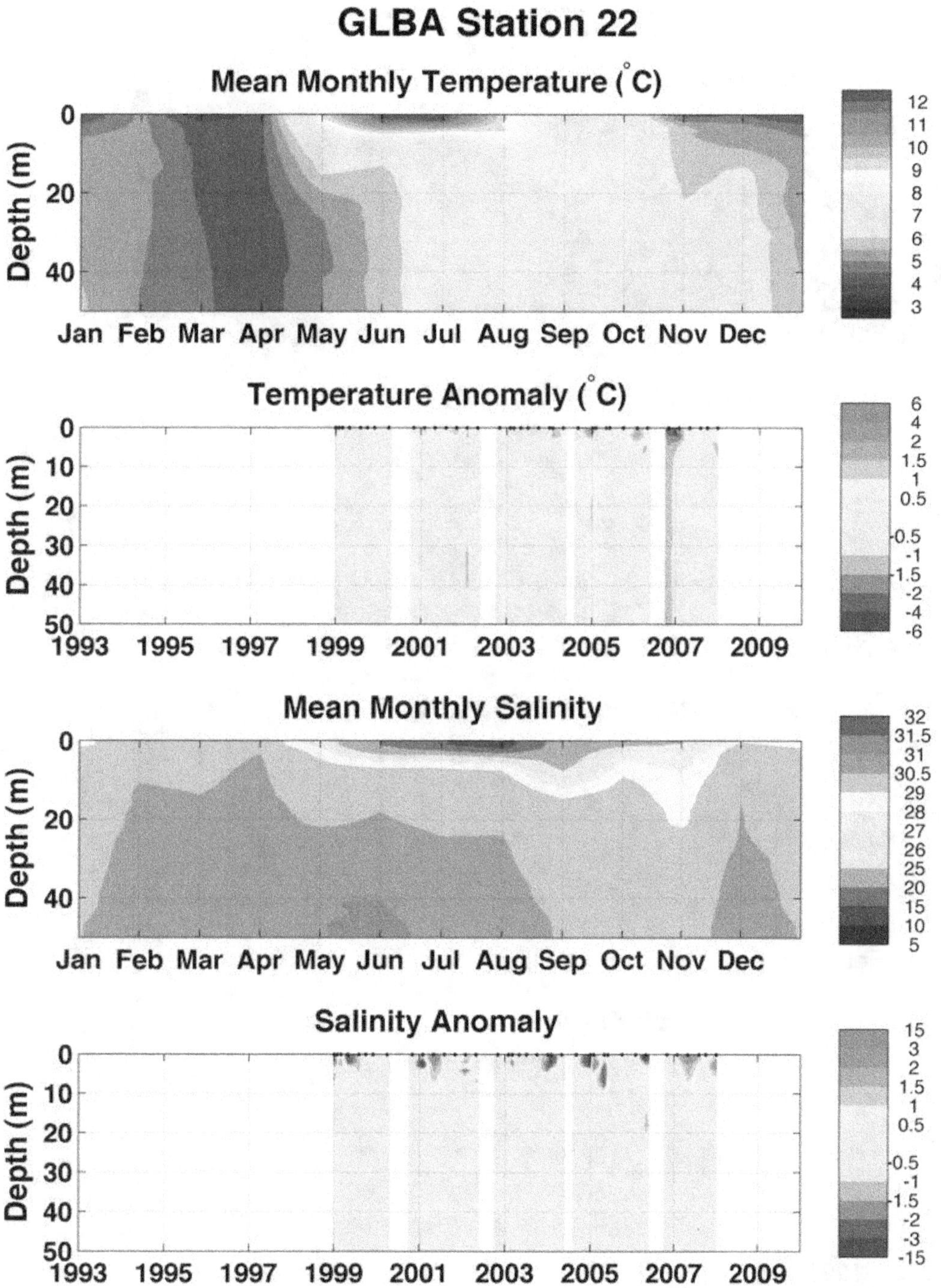

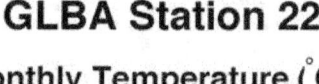

GLBA Station 22

Glacier Bay Oceanographic Monitoring Program Analysis of Observations, 1993–2009
Appendix A. Time-depth sections of temperature and salinity seasonal cycle and anomalies for the period of record over the upper 50 m.

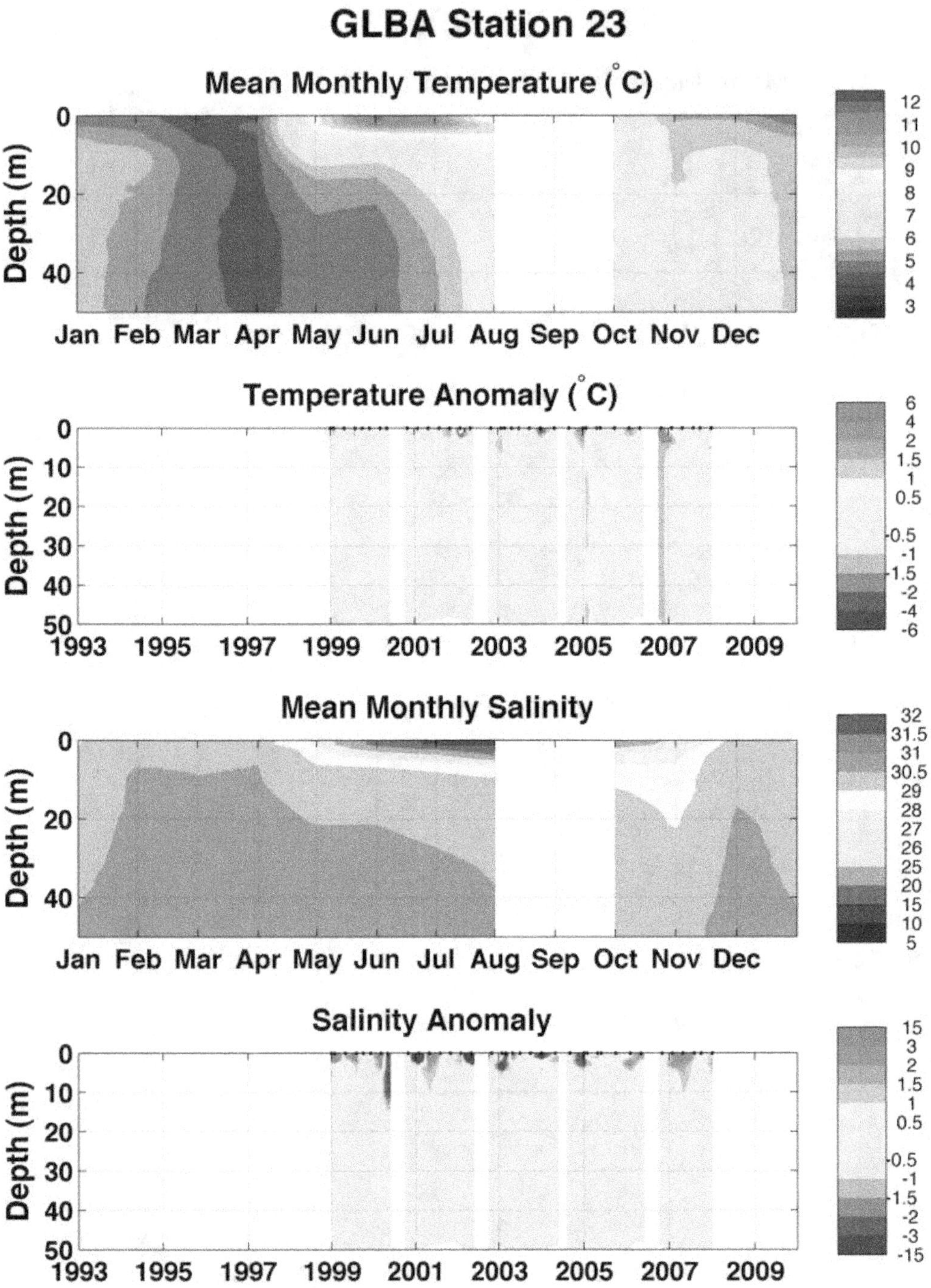

GLBA Station 23

Appendix B. Time-depth sections of temperature and salinity seasonal cycle and anomalies for the period of record over the whole water column.

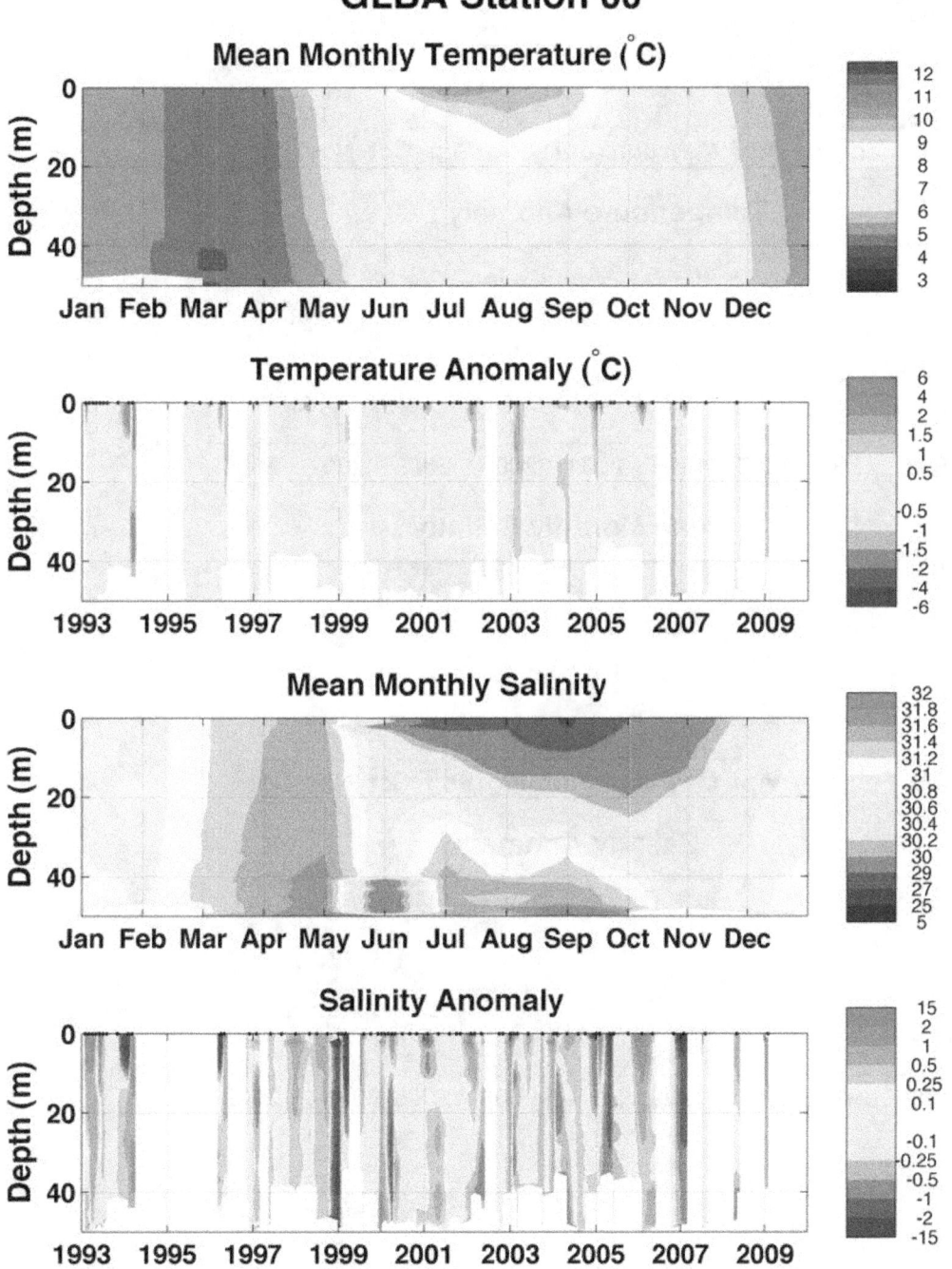

Glacier Bay Oceanographic Monitoring Program Analysis of Observations, 1993–2009
Appendix B. Time-depth sections of temperature and salinity seasonal cycle and anomalies for the period of record over the whole water column.

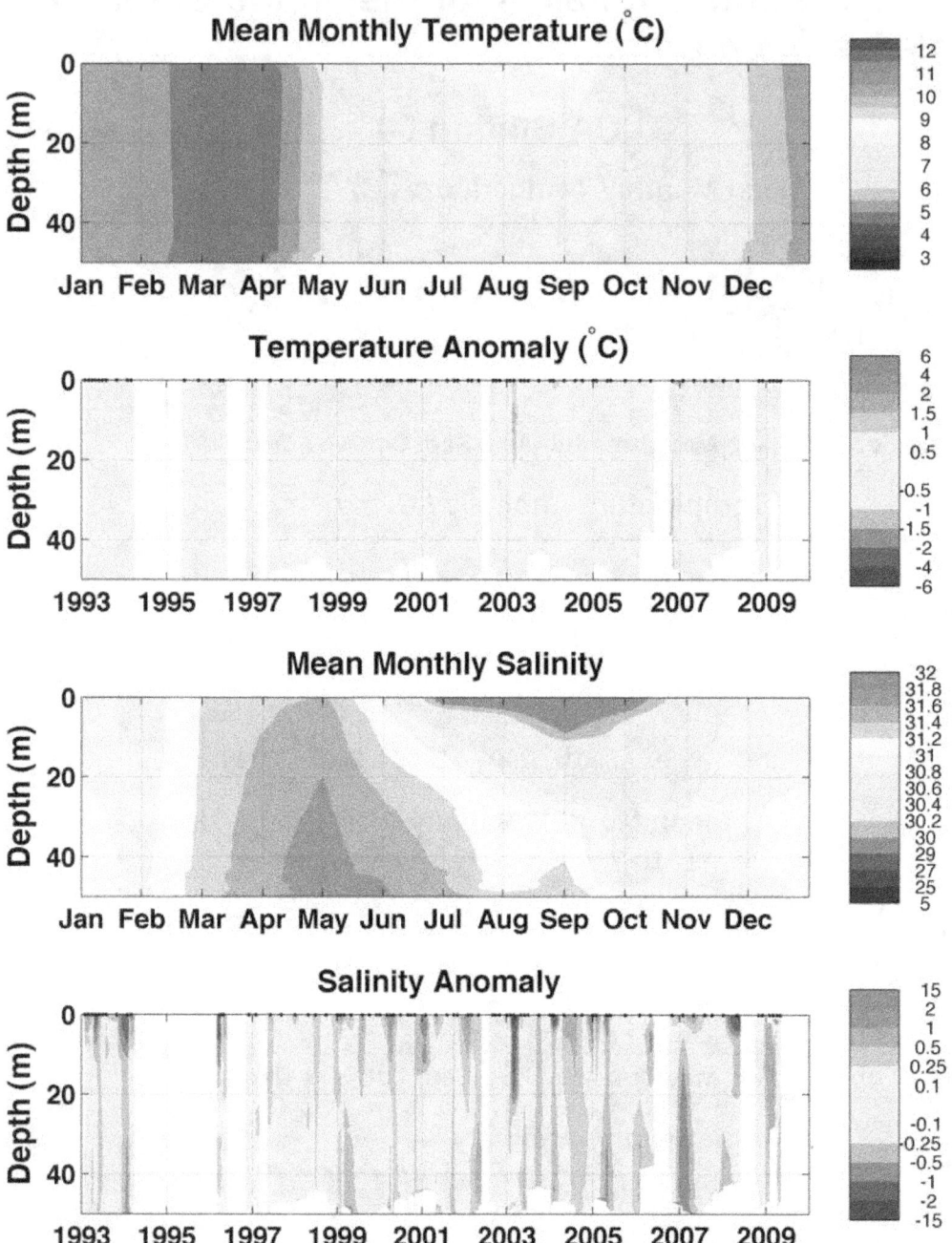

Glacier Bay Oceanographic Monitoring Program Analysis of Observations, 1993–2009
Appendix B. Time-depth sections of temperature and salinity seasonal cycle and anomalies for the period of record over the whole water column.

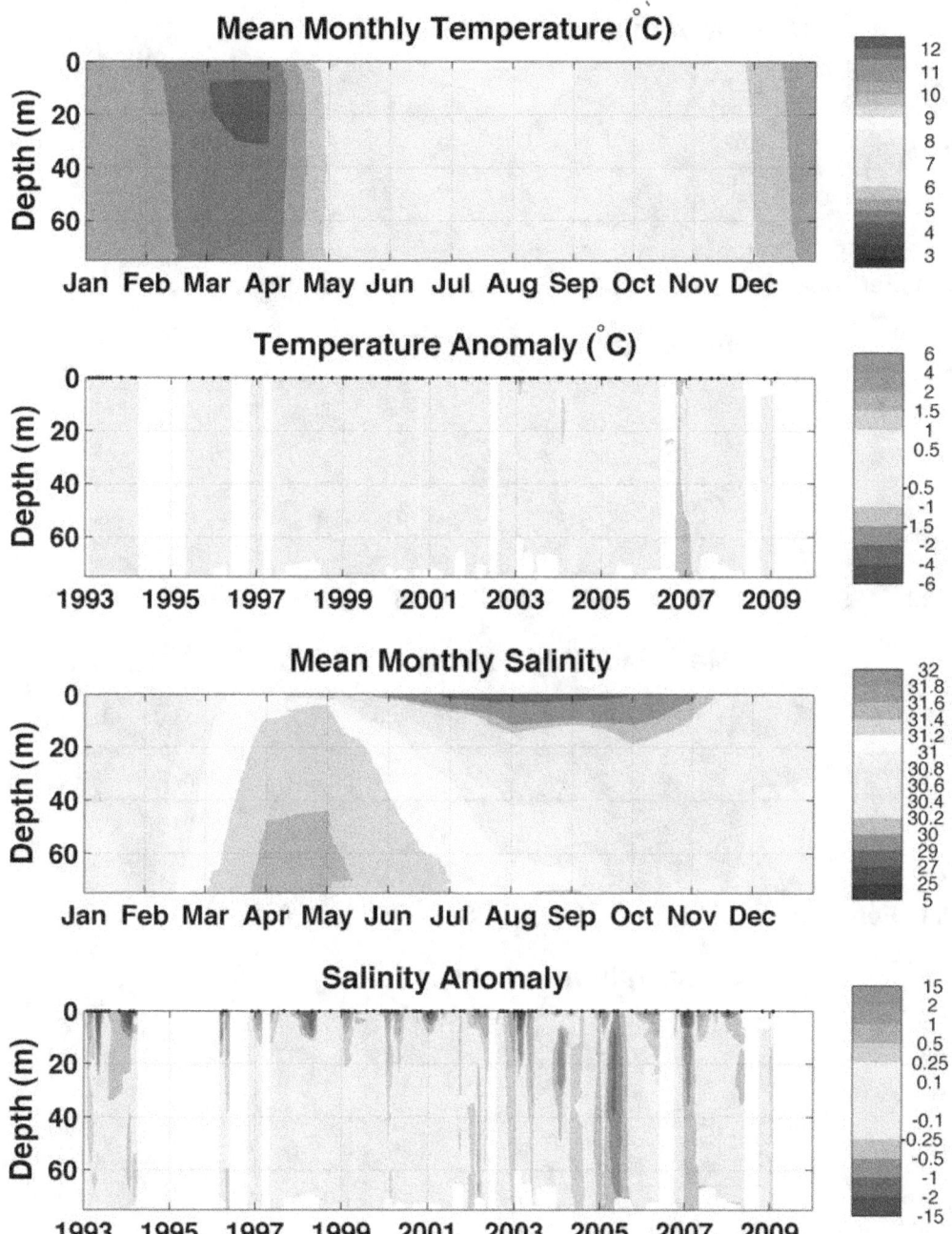

Glacier Bay Oceanographic Monitoring Program Analysis of Observations, 1993–2009
Appendix B. Time-depth sections of temperature and salinity seasonal cycle and anomalies for the period of record over the whole water column.

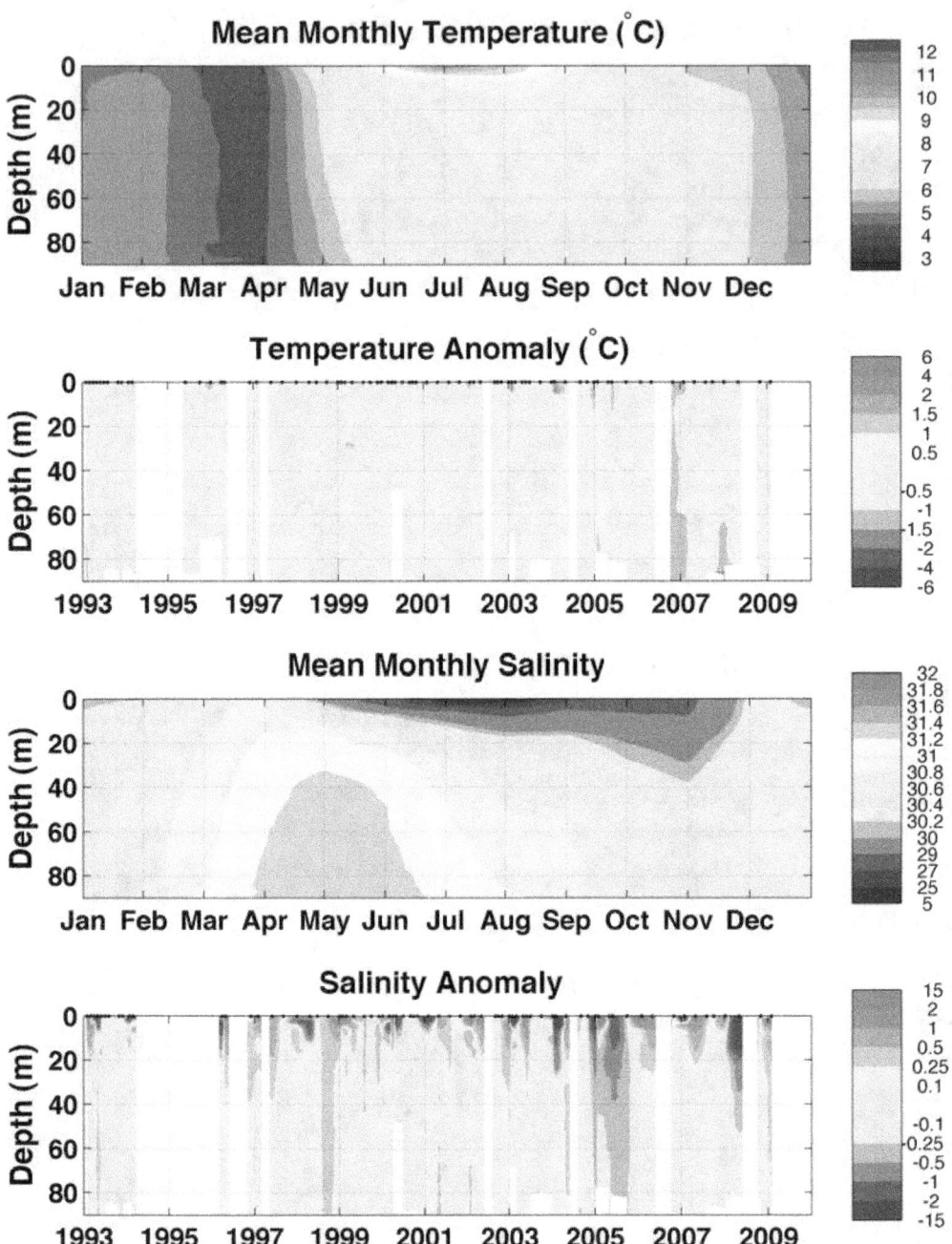

Glacier Bay Oceanographic Monitoring Program Analysis of Observations, 1993–2009
Appendix B. Time-depth sections of temperature and salinity seasonal cycle and anomalies for the period of record over the whole water column.

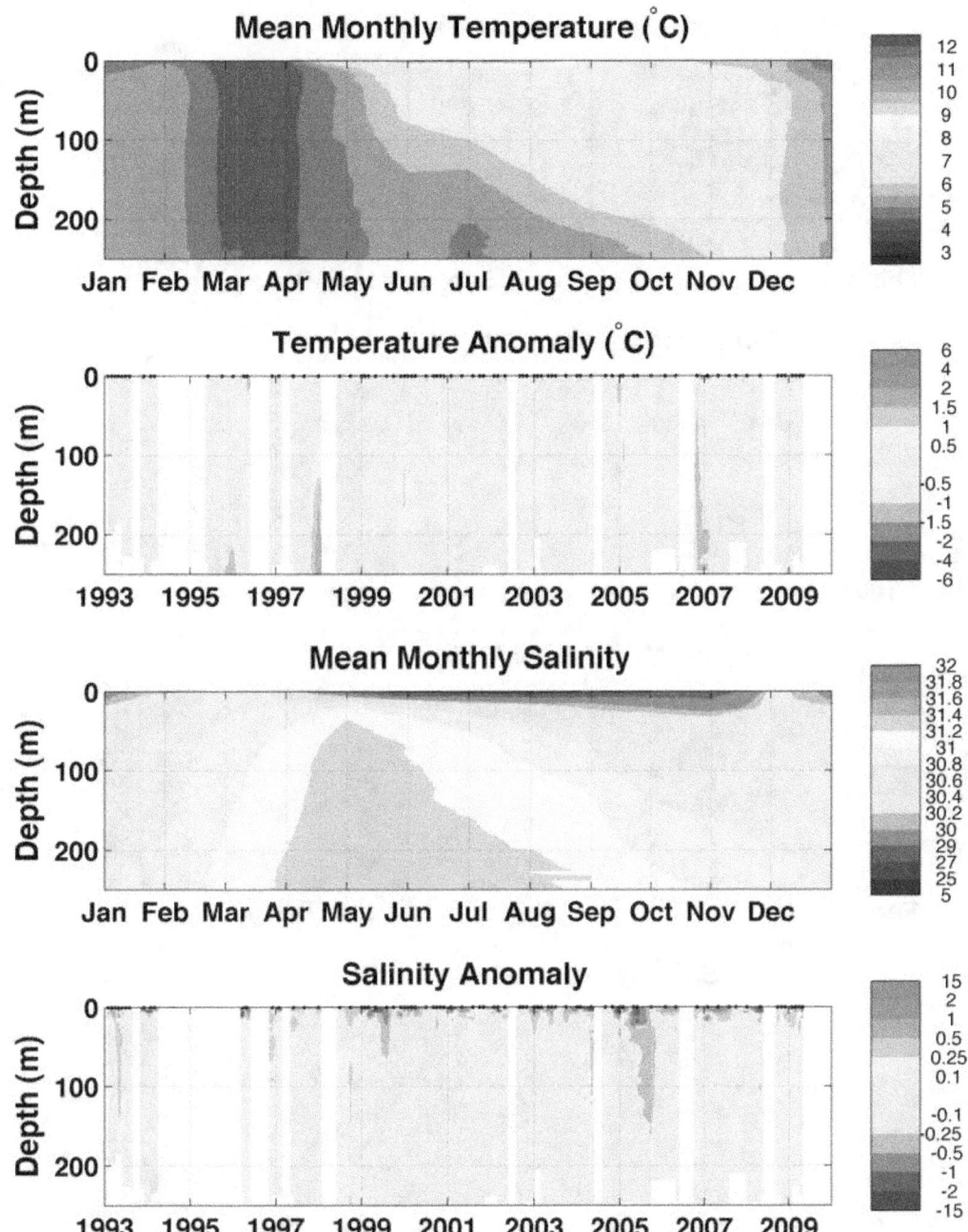

Glacier Bay Oceanographic Monitoring Program Analysis of Observations, 1993–2009
Appendix B. Time-depth sections of temperature and salinity seasonal cycle and anomalies for the period of record over the whole water column.

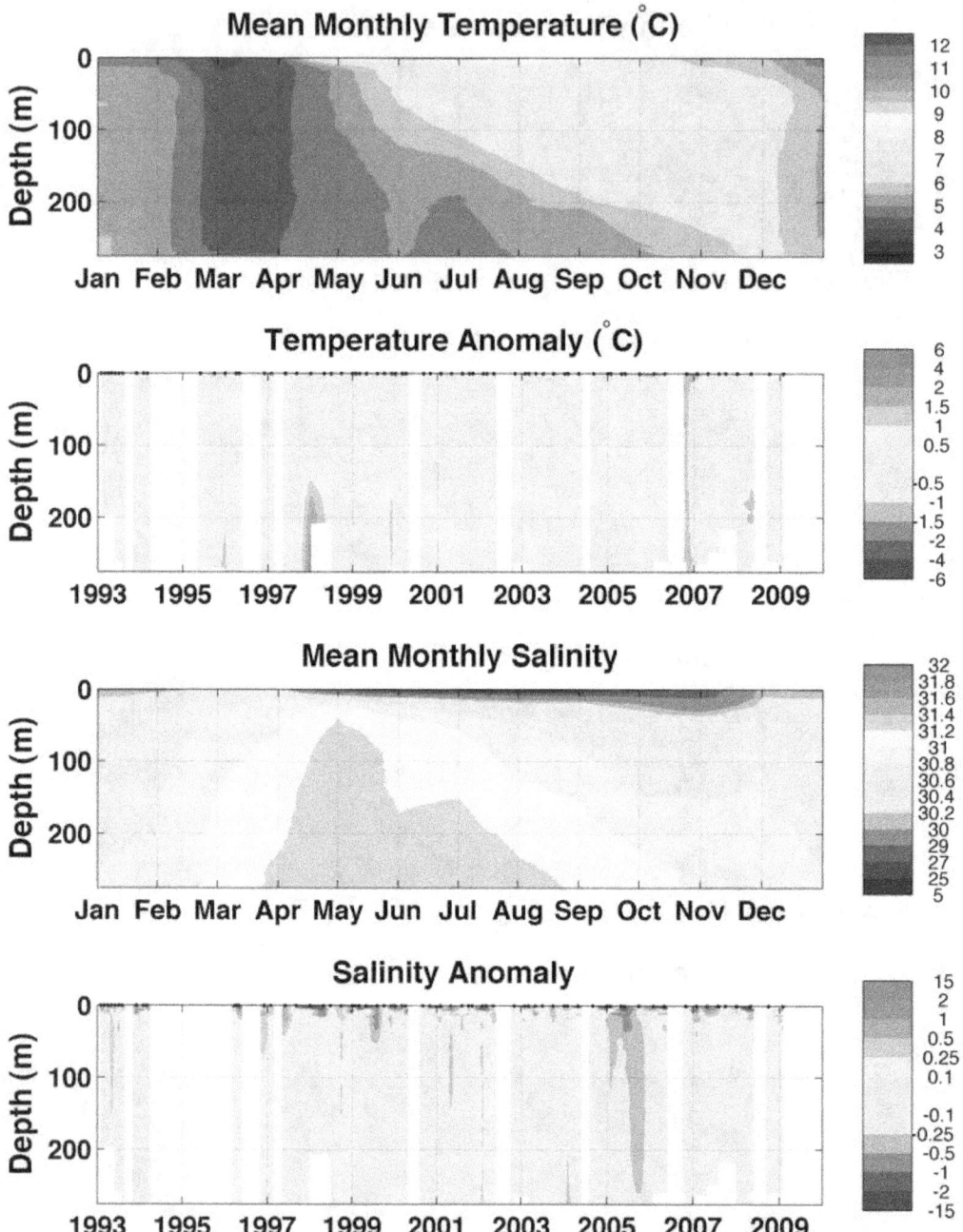

Glacier Bay Oceanographic Monitoring Program Analysis of Observations, 1993–2009
Appendix B. Time-depth sections of temperature and salinity seasonal cycle and anomalies for the period of record over the whole water column.

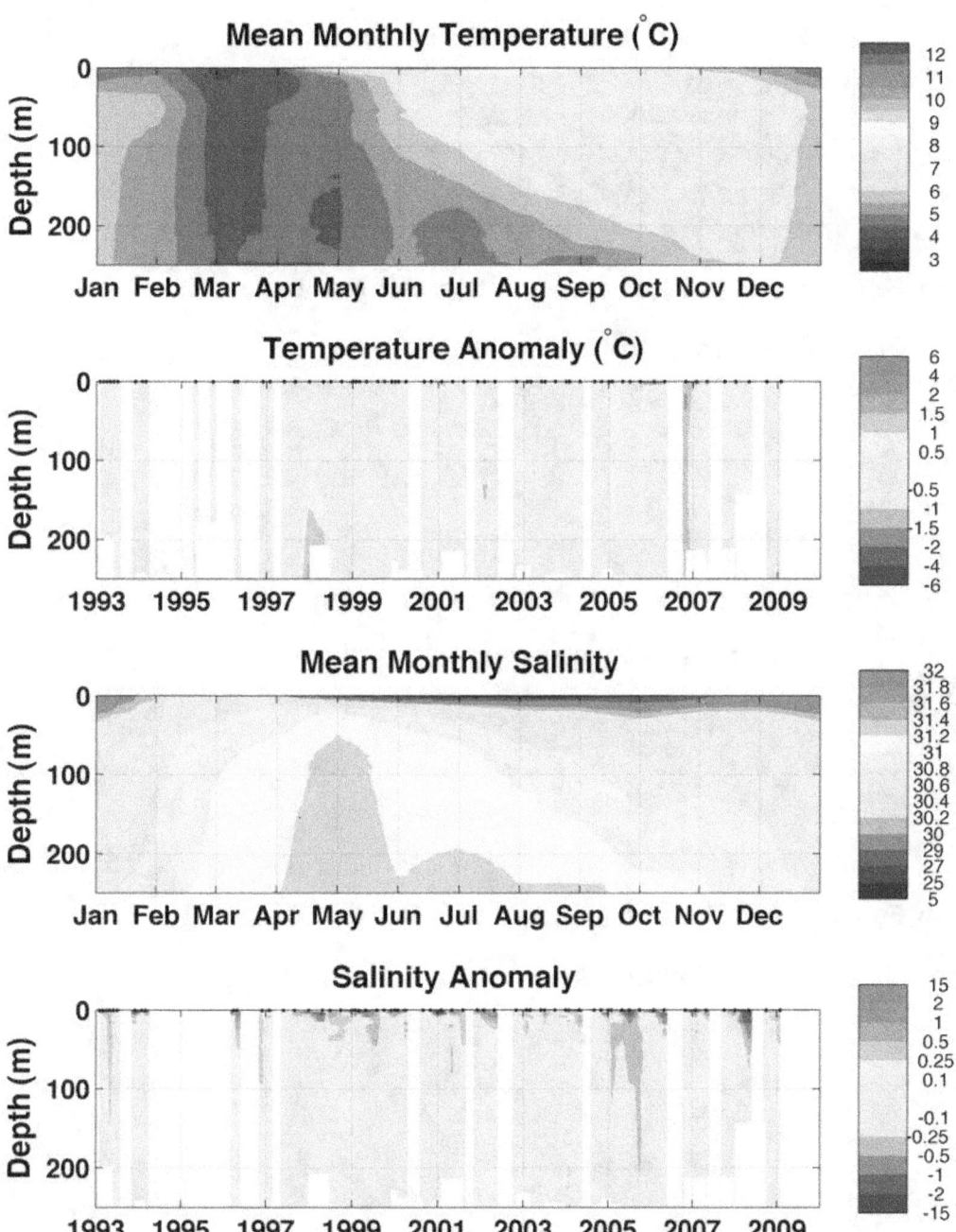

Glacier Bay Oceanographic Monitoring Program Analysis of Observations, 1993–2009
Appendix B. Time-depth sections of temperature and salinity seasonal cycle and anomalies for the period of record over the whole water column.

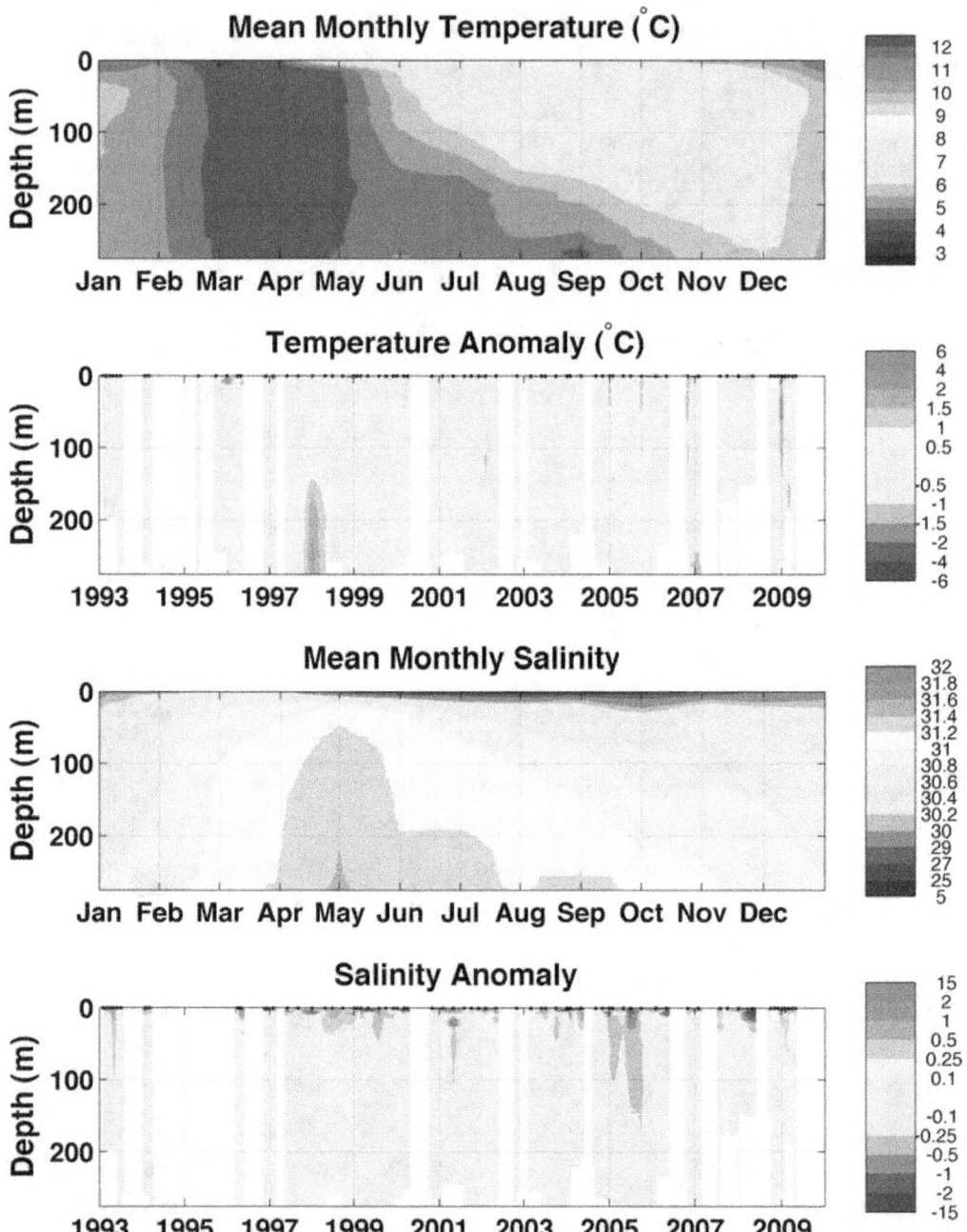

Glacier Bay Oceanographic Monitoring Program Analysis of Observations, 1993–2009
Appendix B. Time-depth sections of temperature and salinity seasonal cycle and anomalies for the period of record over the whole water column.

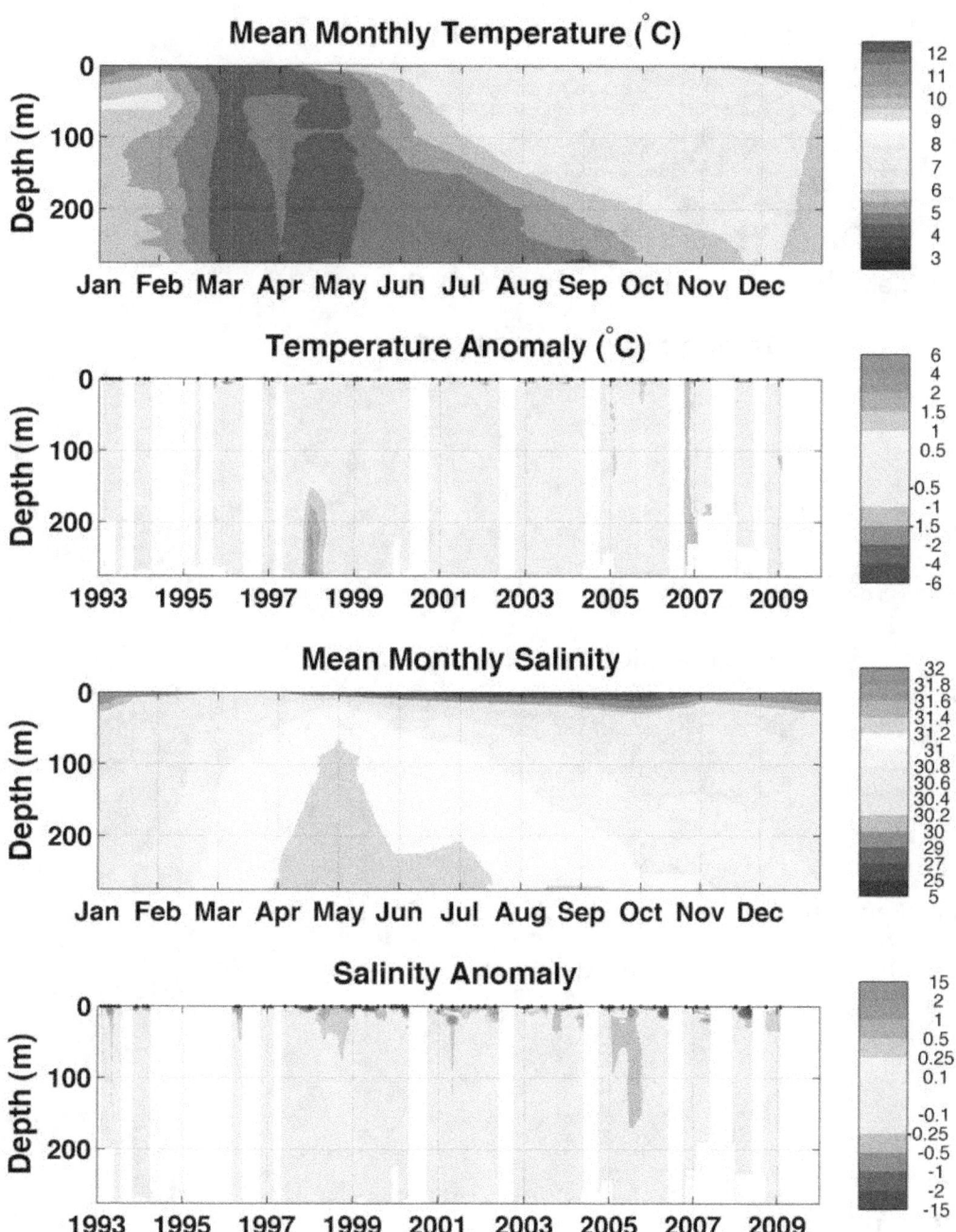

Glacier Bay Oceanographic Monitoring Program Analysis of Observations, 1993–2009
Appendix B. Time-depth sections of temperature and salinity seasonal cycle and anomalies for the period of record over the whole water column.

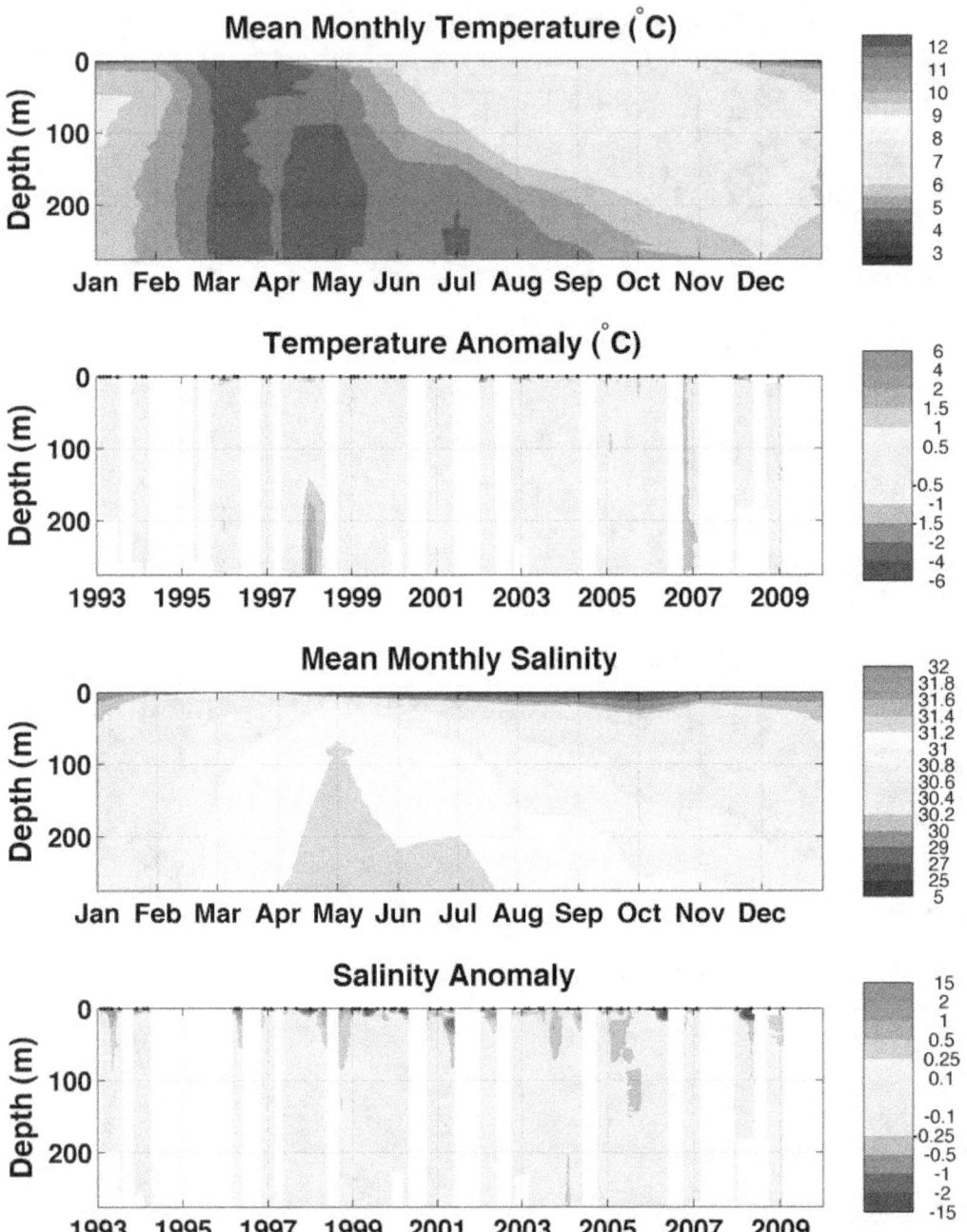

Glacier Bay Oceanographic Monitoring Program Analysis of Observations, 1993–2009
Appendix B. Time-depth sections of temperature and salinity seasonal cycle and anomalies for the period of record over the whole water column.

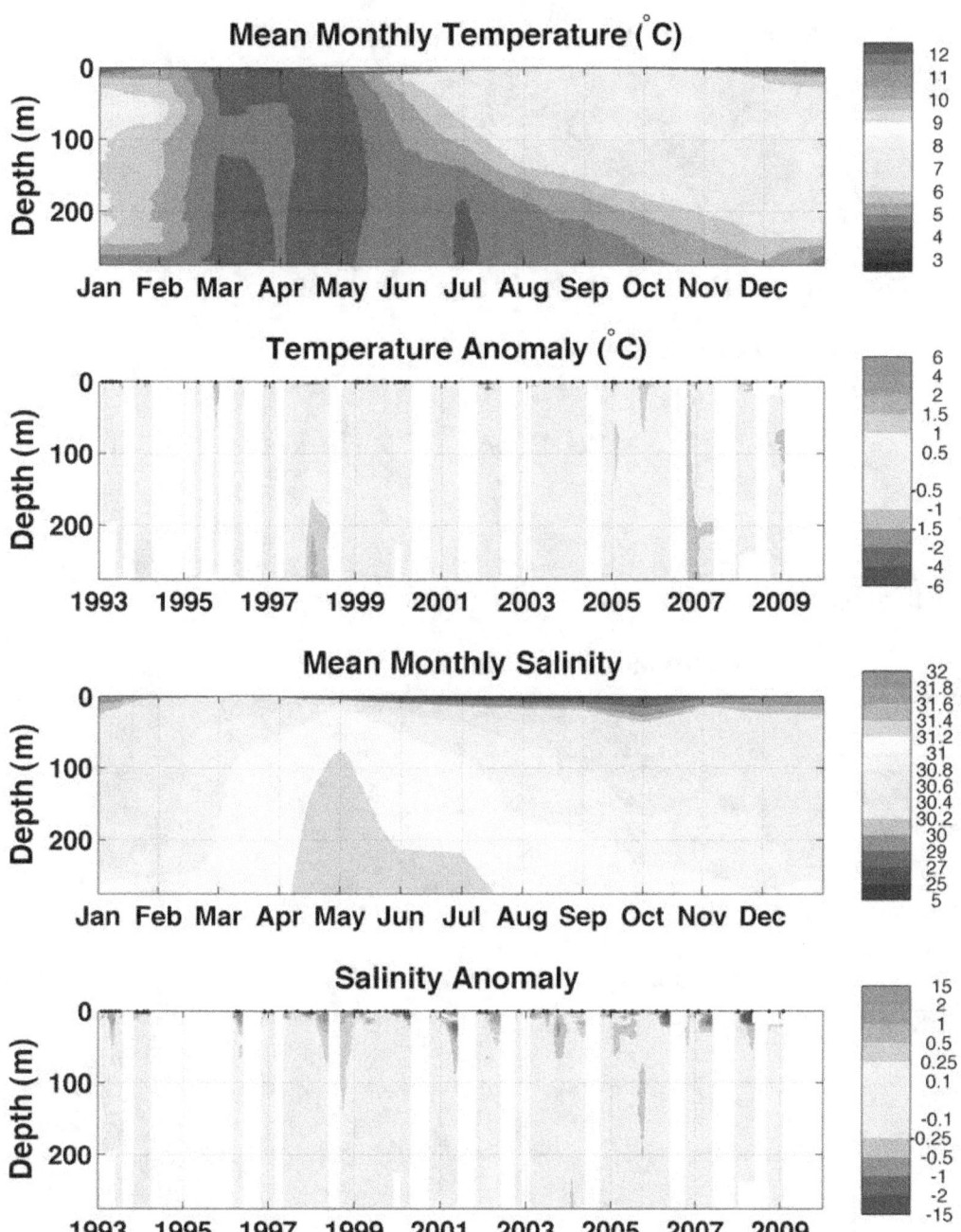

Glacier Bay Oceanographic Monitoring Program Analysis of Observations, 1993–2009
Appendix B. Time-depth sections of temperature and salinity seasonal cycle and anomalies for the period of record over the whole water column.

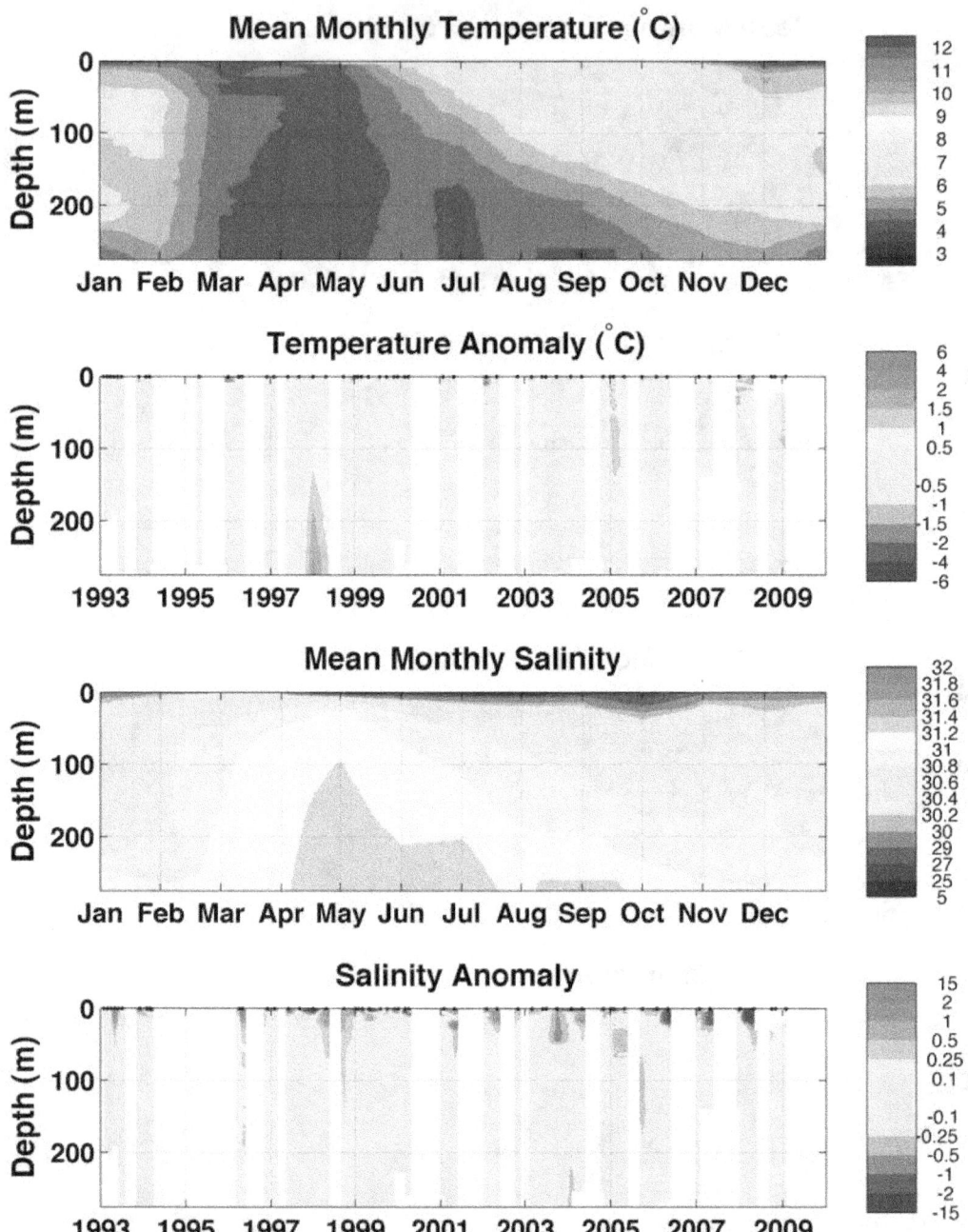

Glacier Bay Oceanographic Monitoring Program Analysis of Observations, 1993–2009
Appendix B. Time-depth sections of temperature and salinity seasonal cycle and anomalies for the period of record over the whole water column.

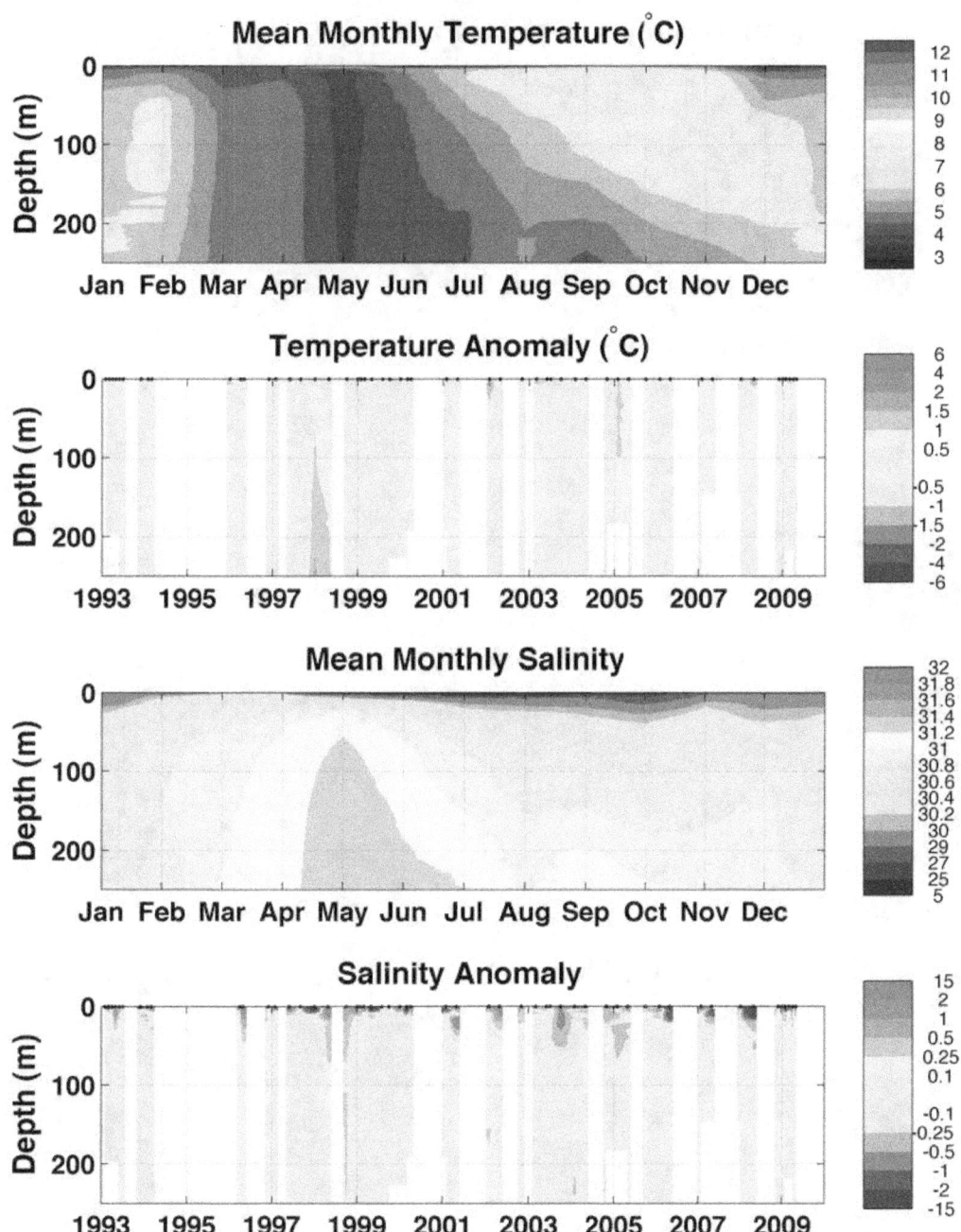

Glacier Bay Oceanographic Monitoring Program Analysis of Observations, 1993–2009
Appendix B. Time-depth sections of temperature and salinity seasonal cycle and anomalies for the period of record over the whole water column.

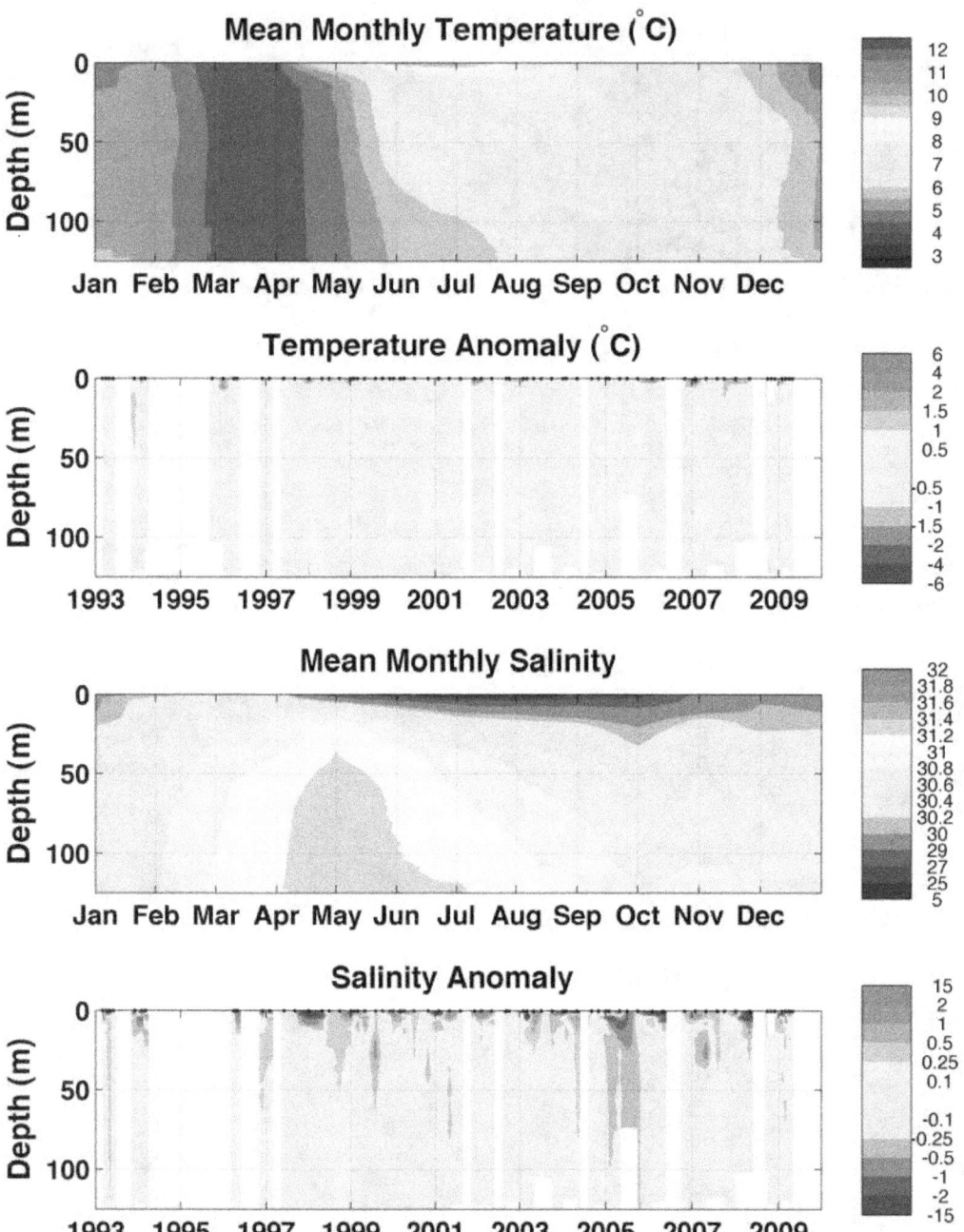

Glacier Bay Oceanographic Monitoring Program Analysis of Observations, 1993–2009
Appendix B. Time-depth sections of temperature and salinity seasonal cycle and anomalies for the period of record over the whole water column.

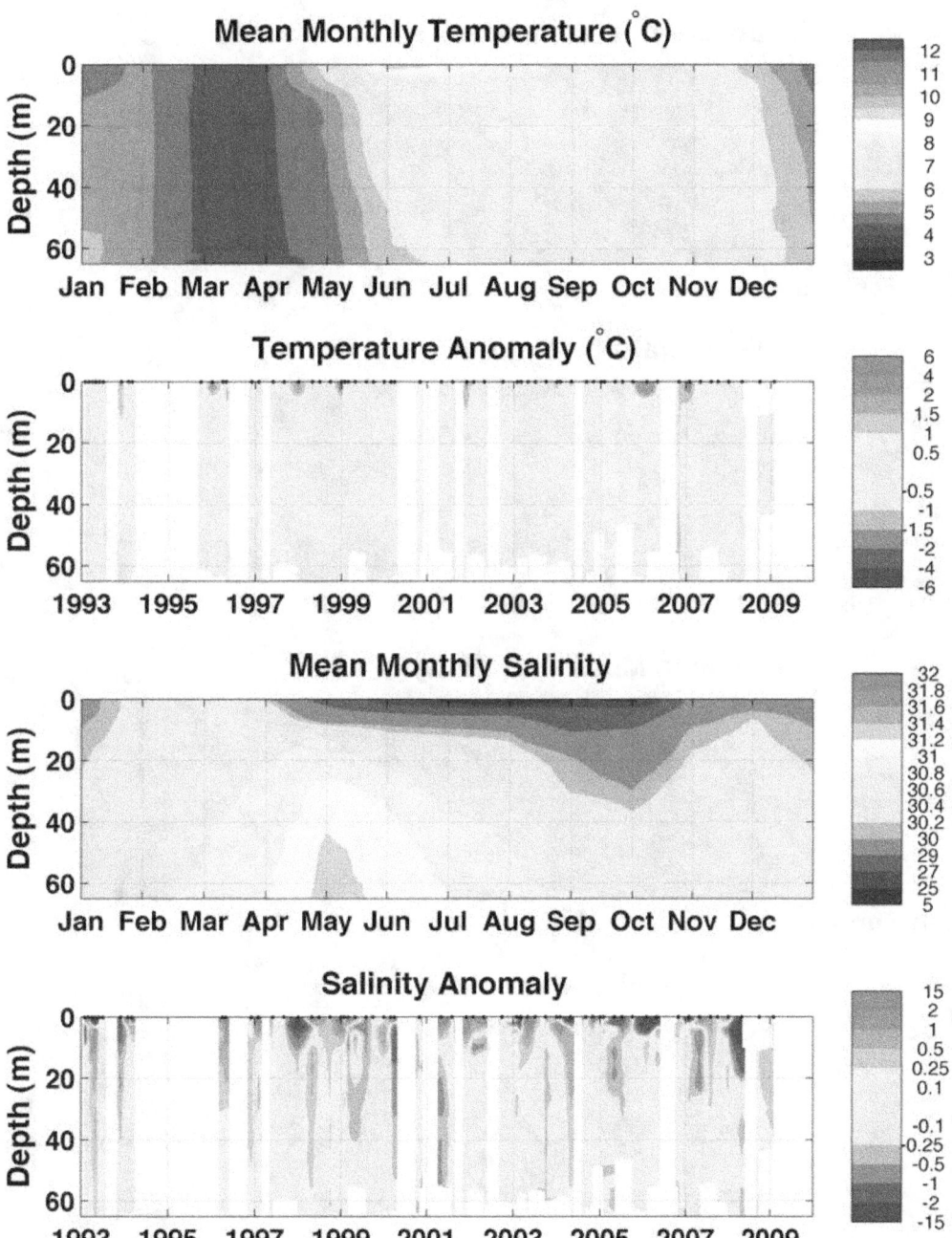

Glacier Bay Oceanographic Monitoring Program Analysis of Observations, 1993–2009
Appendix B. Time-depth sections of temperature and salinity seasonal cycle and anomalies for the period of record over the whole water column.

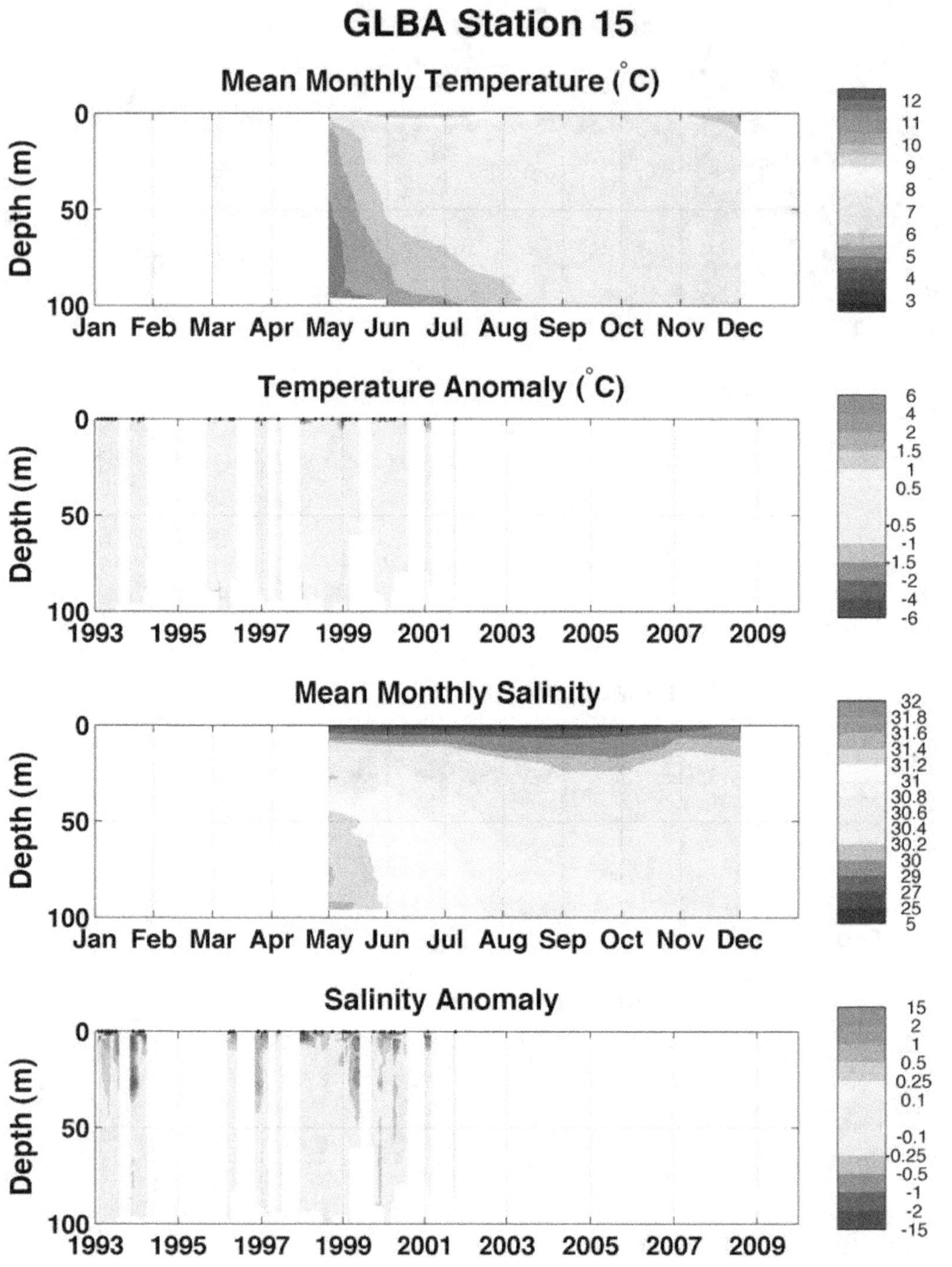

GLBA Station 15

Glacier Bay Oceanographic Monitoring Program Analysis of Observations, 1993–2009
Appendix B. Time-depth sections of temperature and salinity seasonal cycle and anomalies for the period of record over the whole water column.

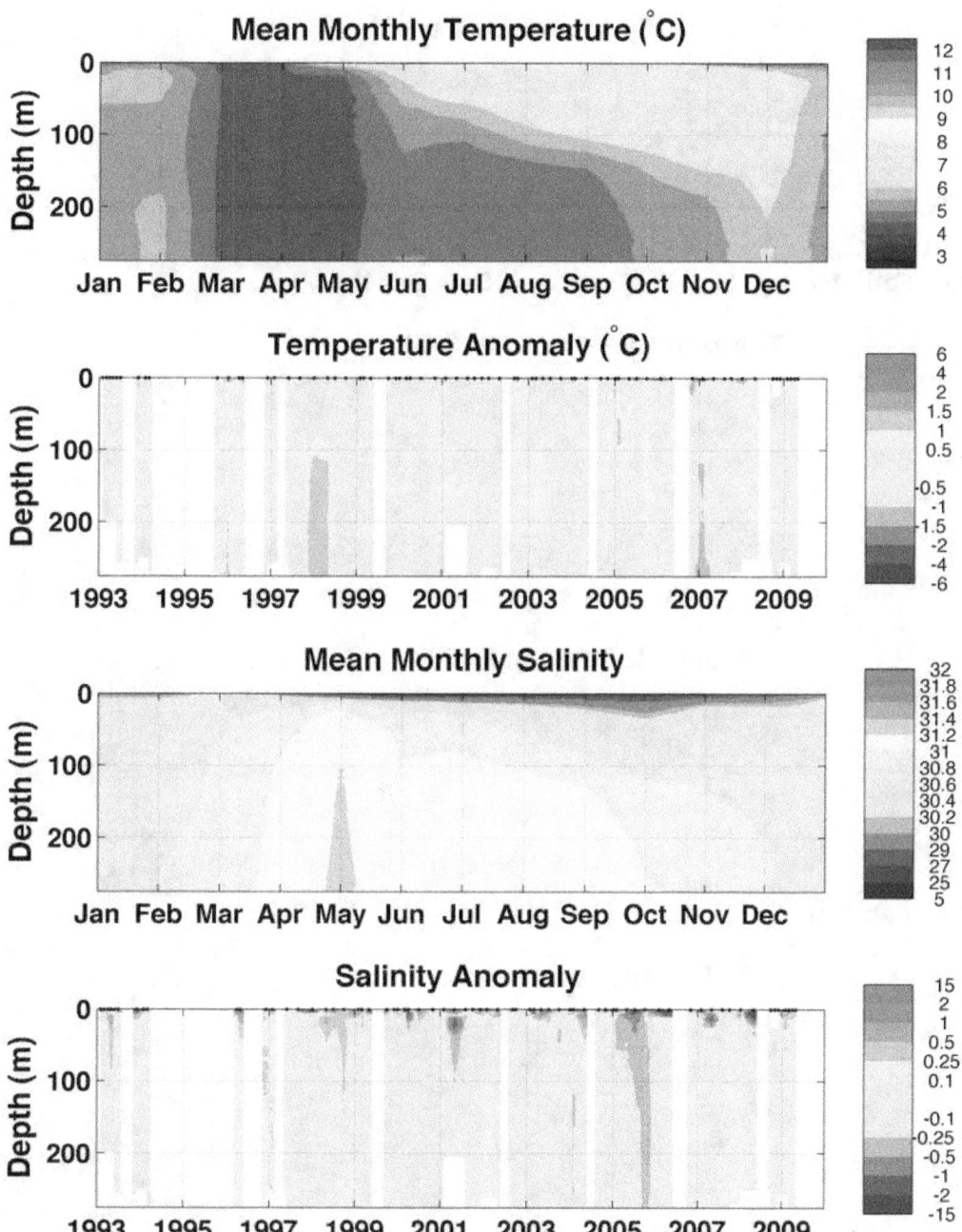

Glacier Bay Oceanographic Monitoring Program Analysis of Observations, 1993–2009
Appendix B. Time-depth sections of temperature and salinity seasonal cycle and anomalies for the period of record over the whole water column.

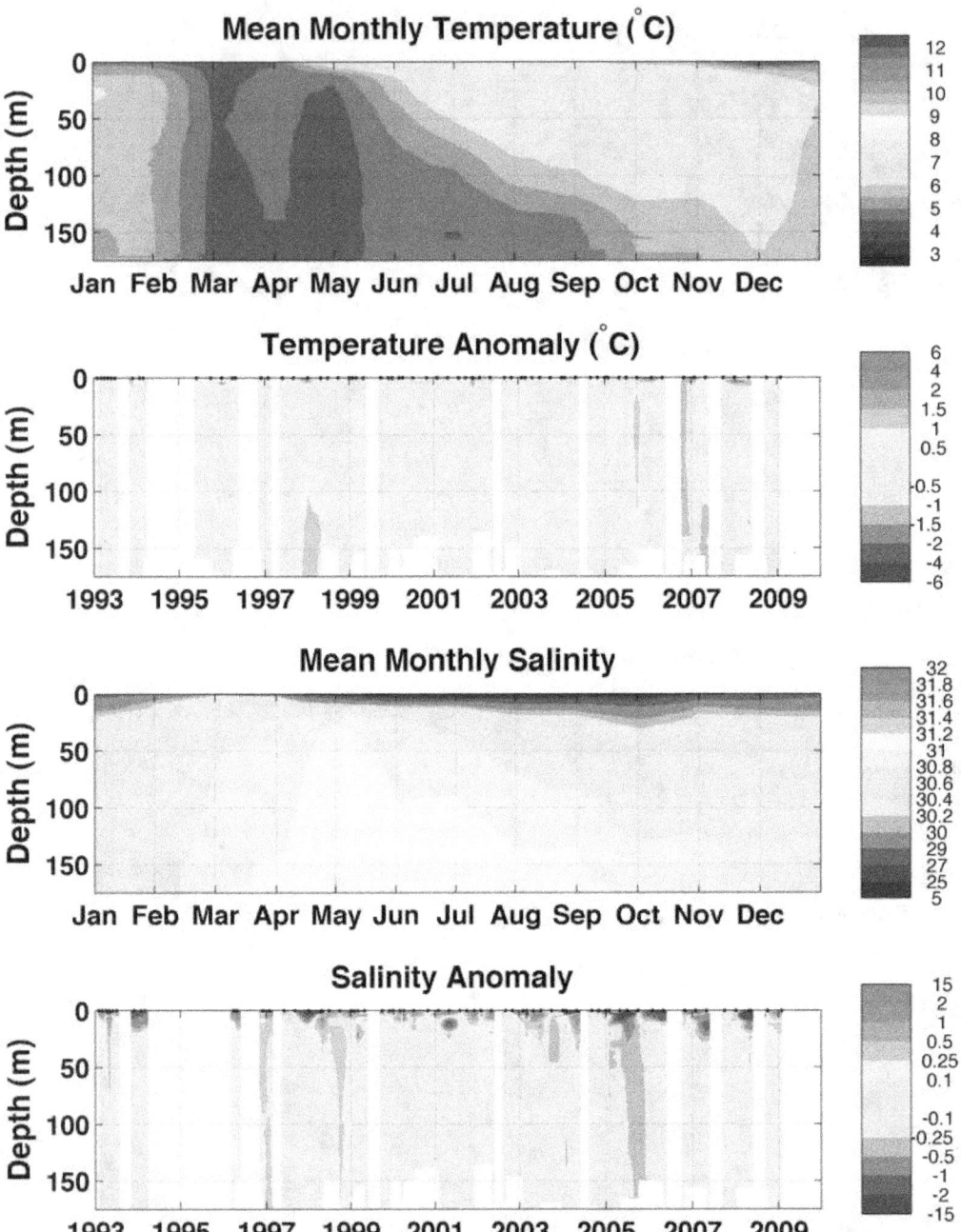

Glacier Bay Oceanographic Monitoring Program Analysis of Observations, 1993–2009
Appendix B. Time-depth sections of temperature and salinity seasonal cycle and anomalies for the period of record over the whole water column.

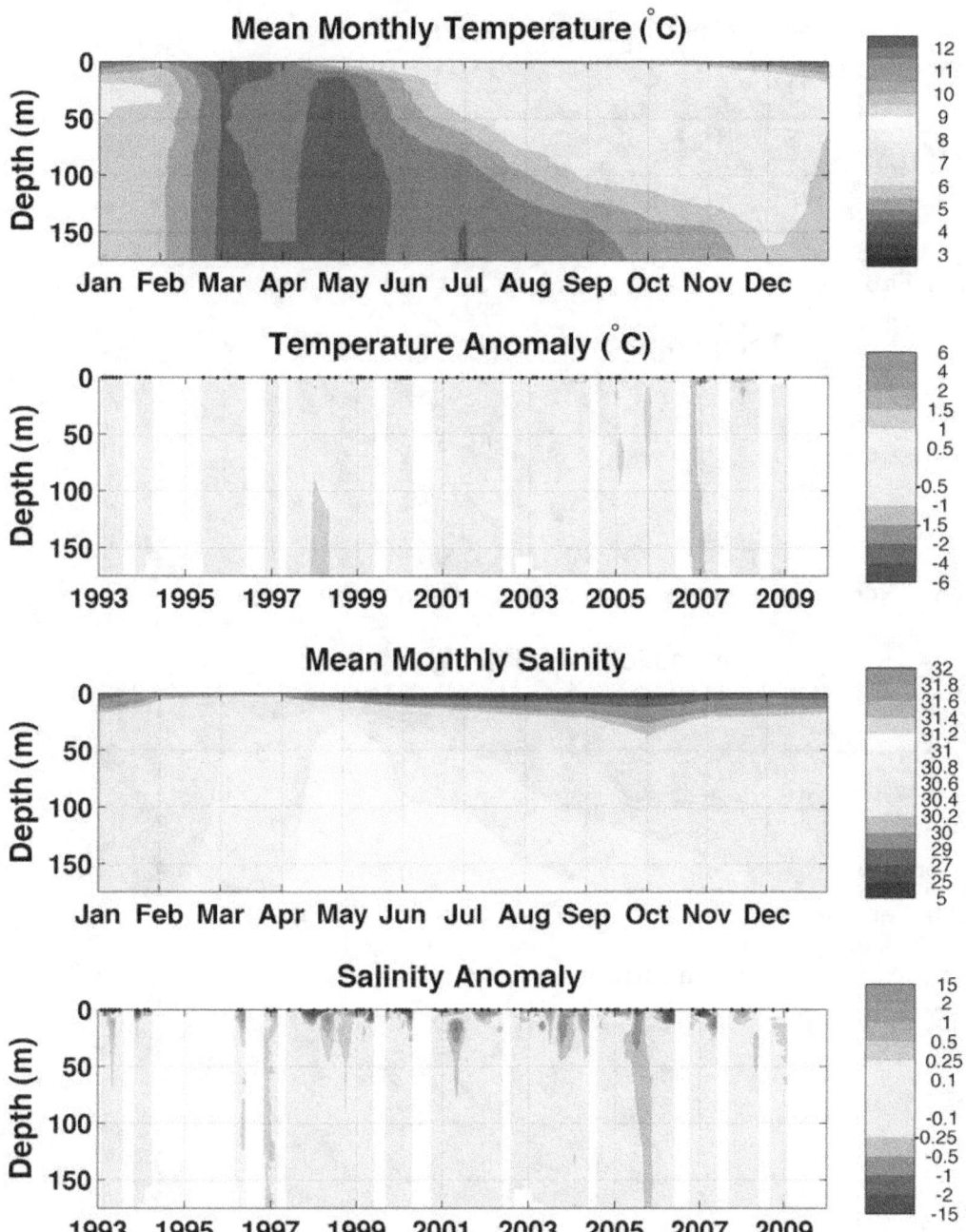

Glacier Bay Oceanographic Monitoring Program Analysis of Observations, 1993–2009
Appendix B. Time-depth sections of temperature and salinity seasonal cycle and anomalies for the period of record over the whole water column.

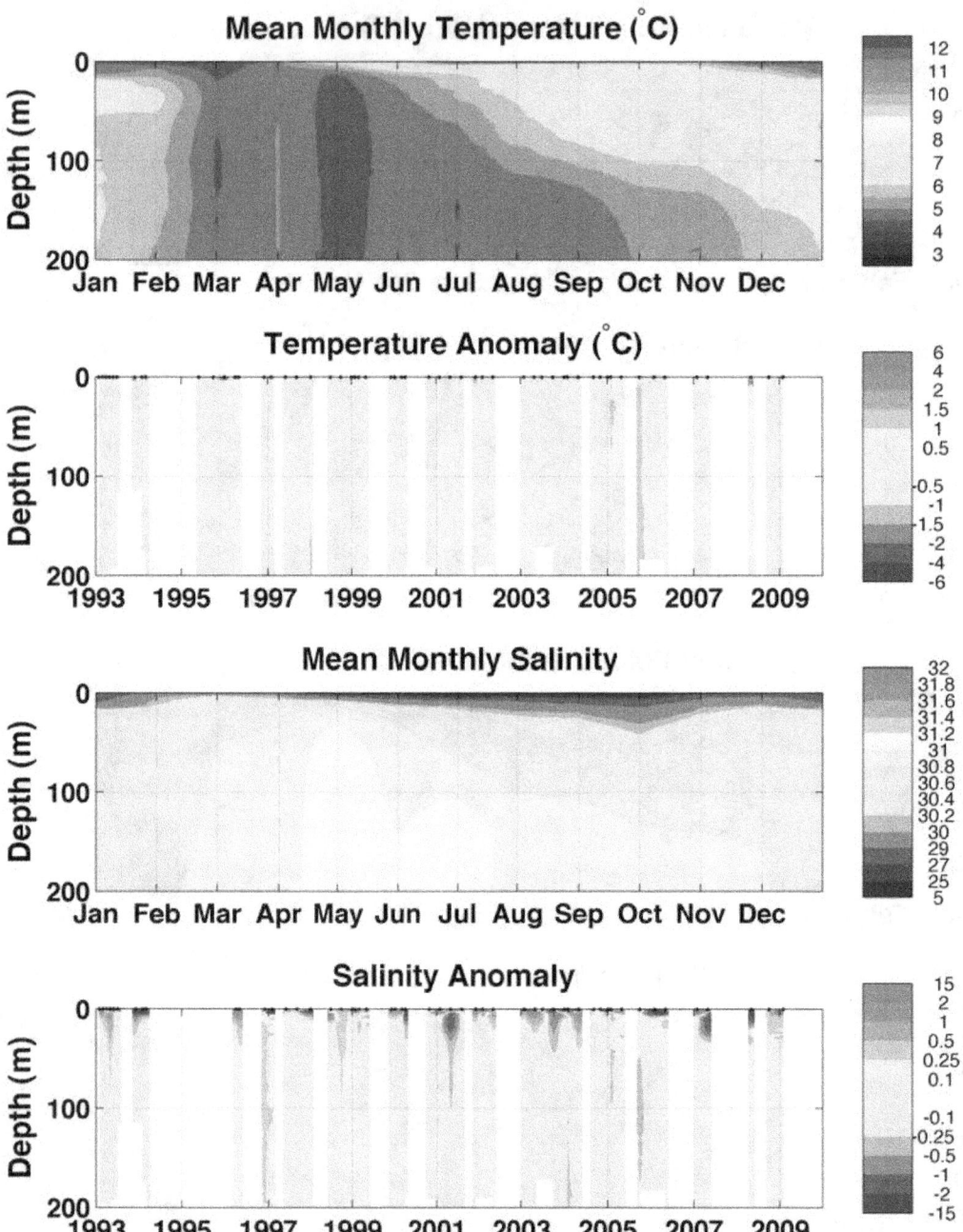

Glacier Bay Oceanographic Monitoring Program Analysis of Observations, 1993–2009
Appendix B. Time-depth sections of temperature and salinity seasonal cycle and anomalies for the period of record over the whole water column.

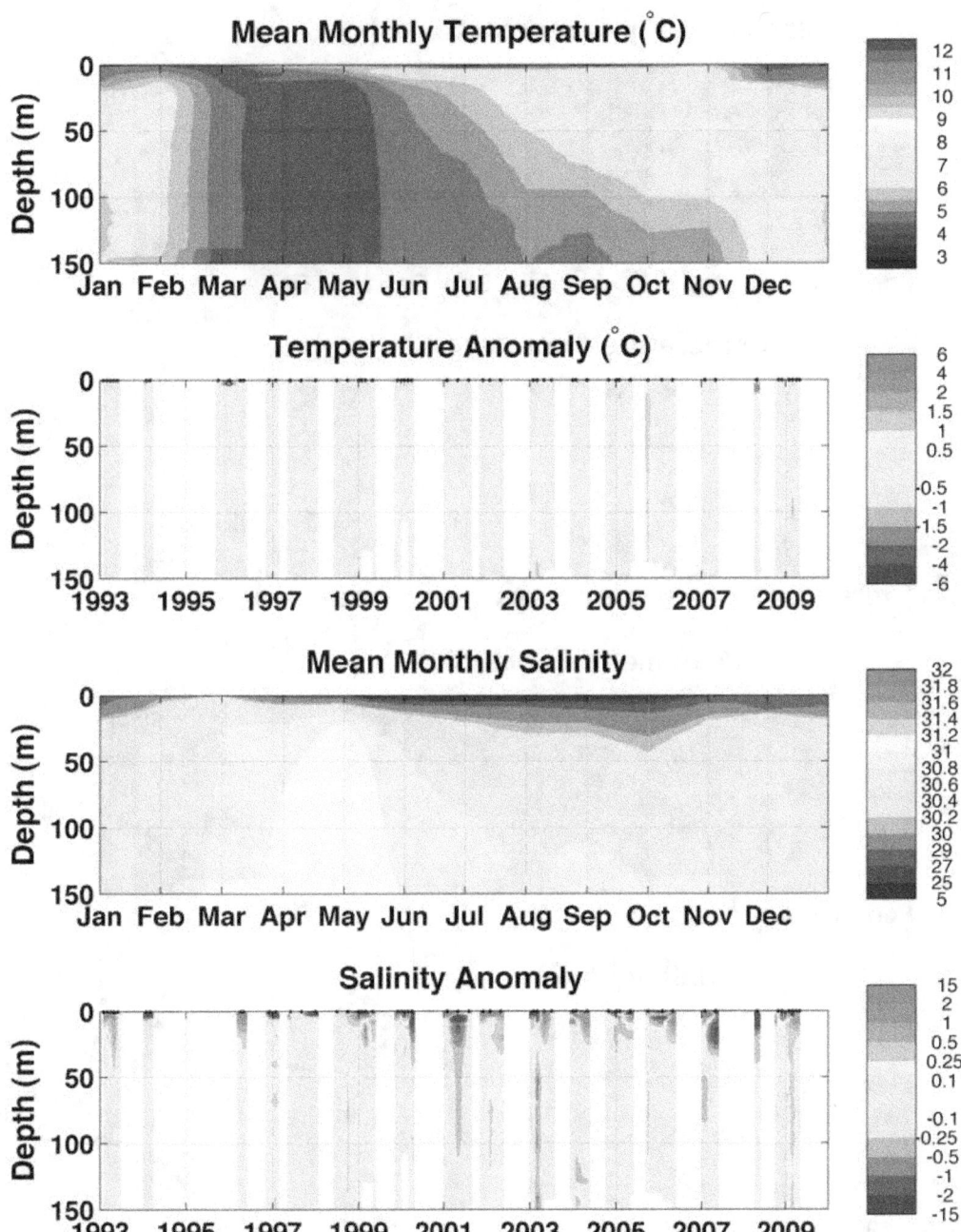

Glacier Bay Oceanographic Monitoring Program Analysis of Observations, 1993–2009
Appendix B. Time-depth sections of temperature and salinity seasonal cycle and anomalies for the period of record over the whole water column.

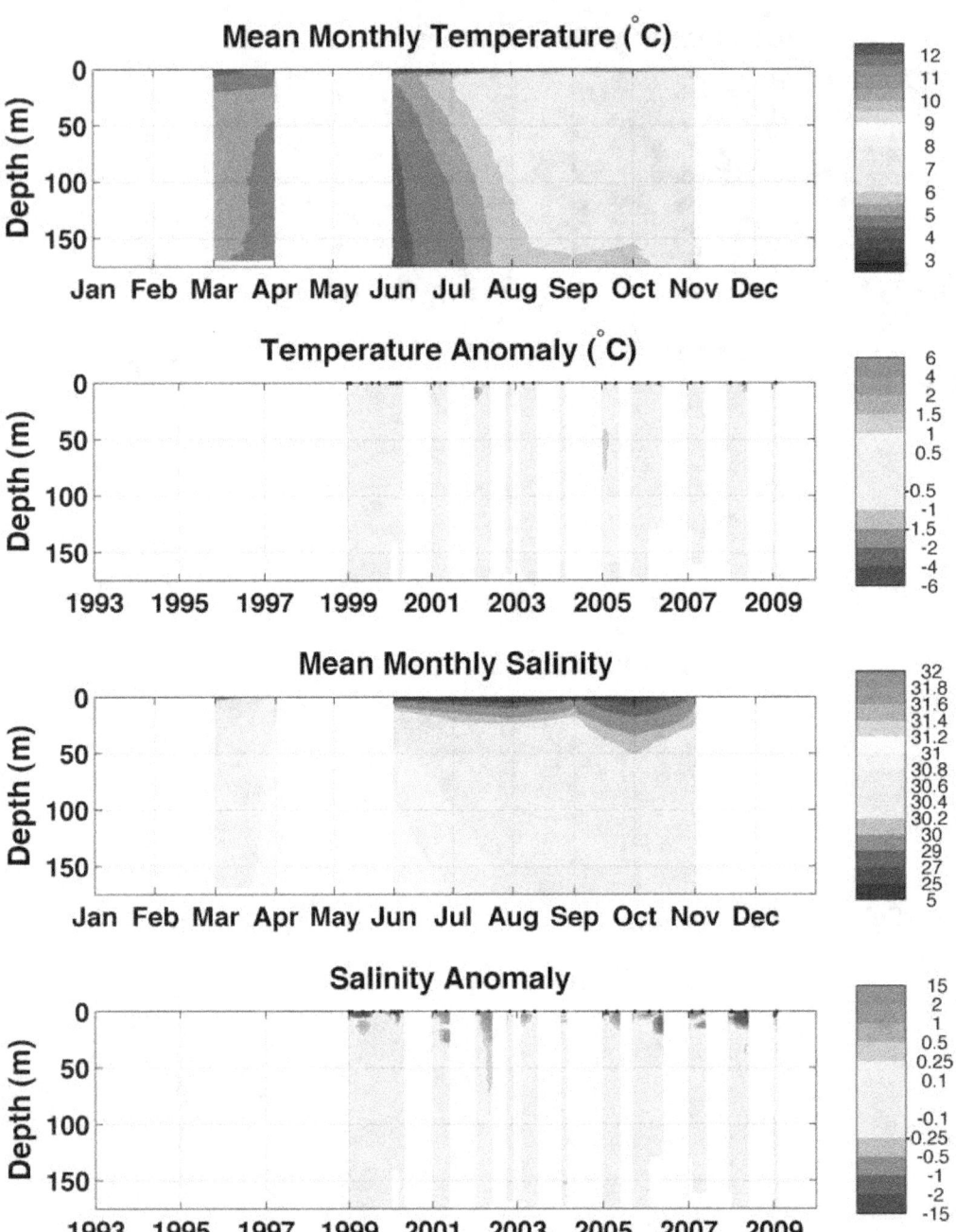

Glacier Bay Oceanographic Monitoring Program Analysis of Observations, 1993–2009
Appendix B. Time-depth sections of temperature and salinity seasonal cycle and anomalies for the period of record over the whole water column.

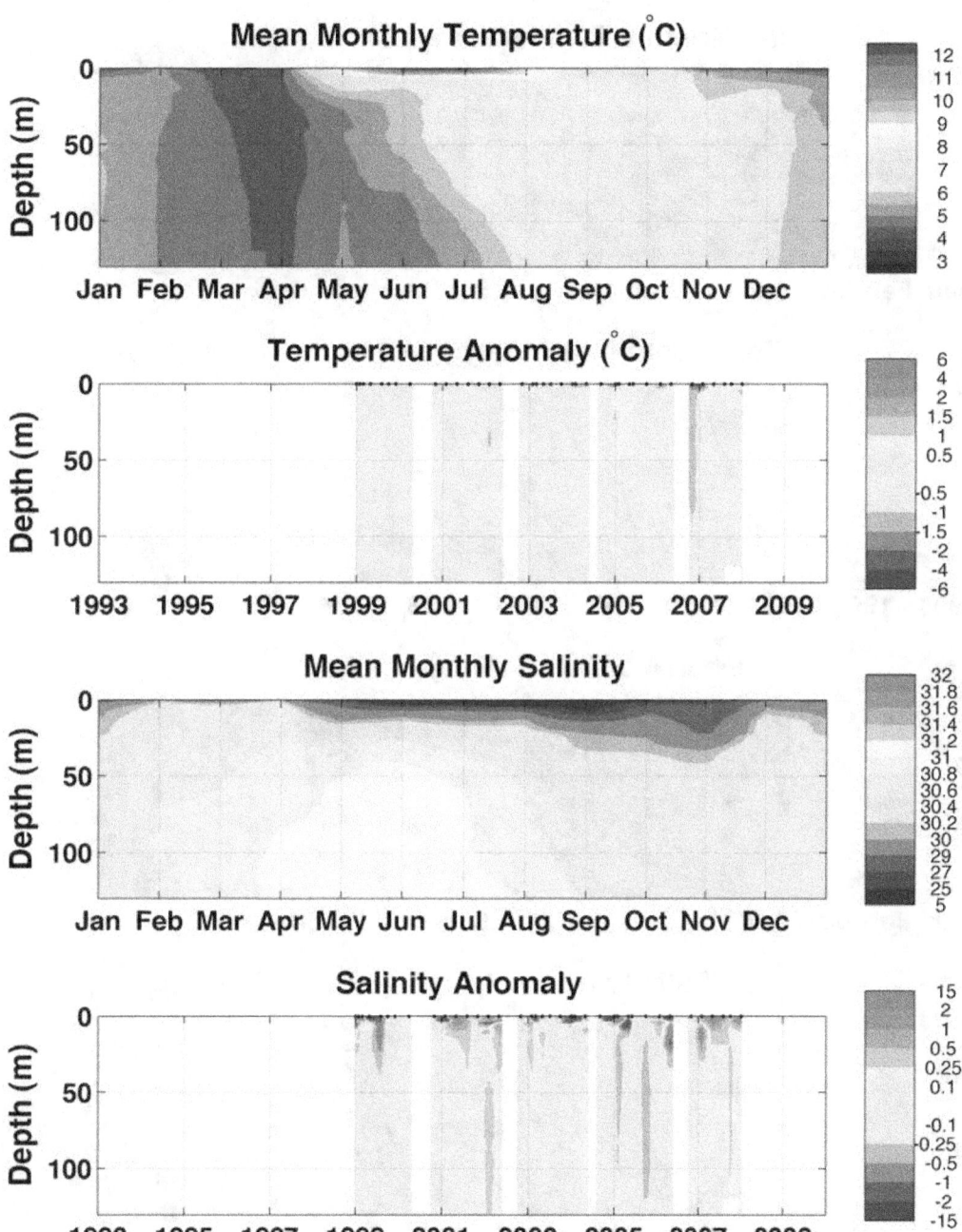

Glacier Bay Oceanographic Monitoring Program Analysis of Observations, 1993–2009
Appendix B. Time-depth sections of temperature and salinity seasonal cycle and anomalies for the period of record over the whole water column.

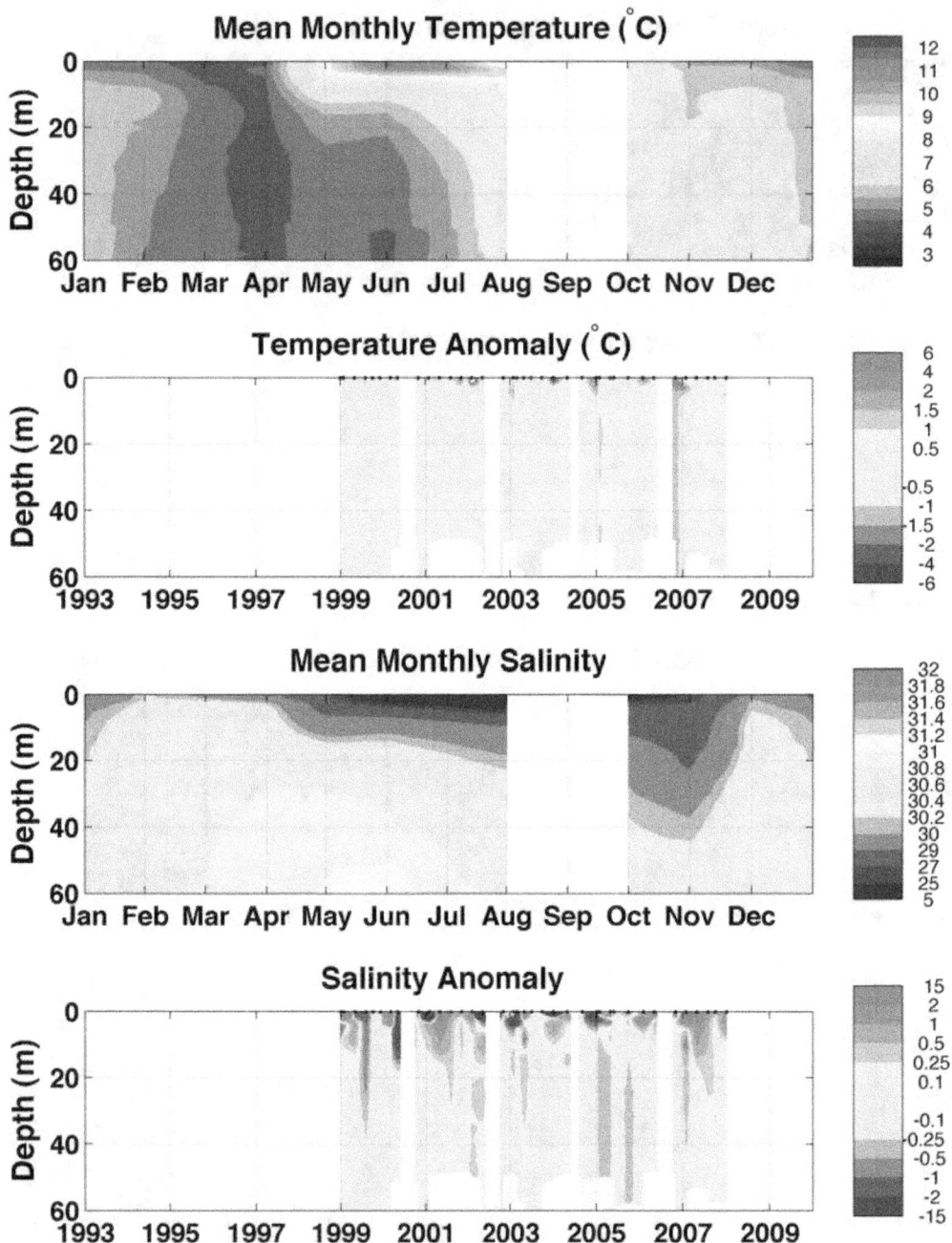

Appendix C. Time series of temperature and salinity anomalies at select depths.

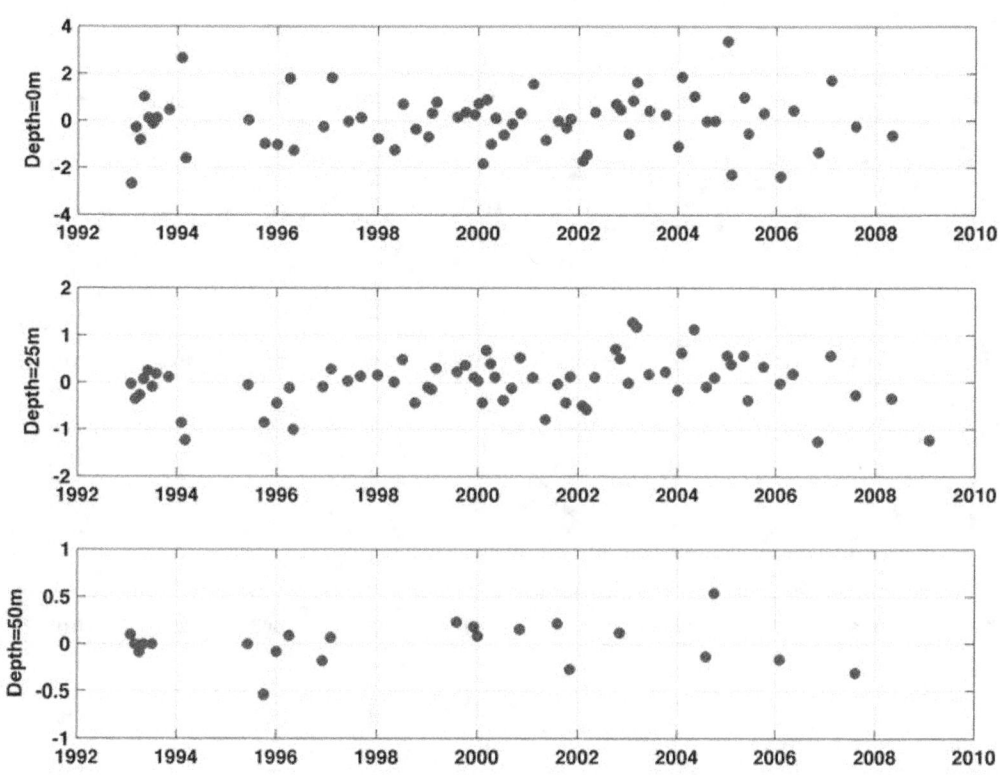

Glacier Bay Oceanographic Monitoring Program Analysis of Observations, 1993–2009
Appendix C. Time series of temperature and salinity anomalies at select depths.

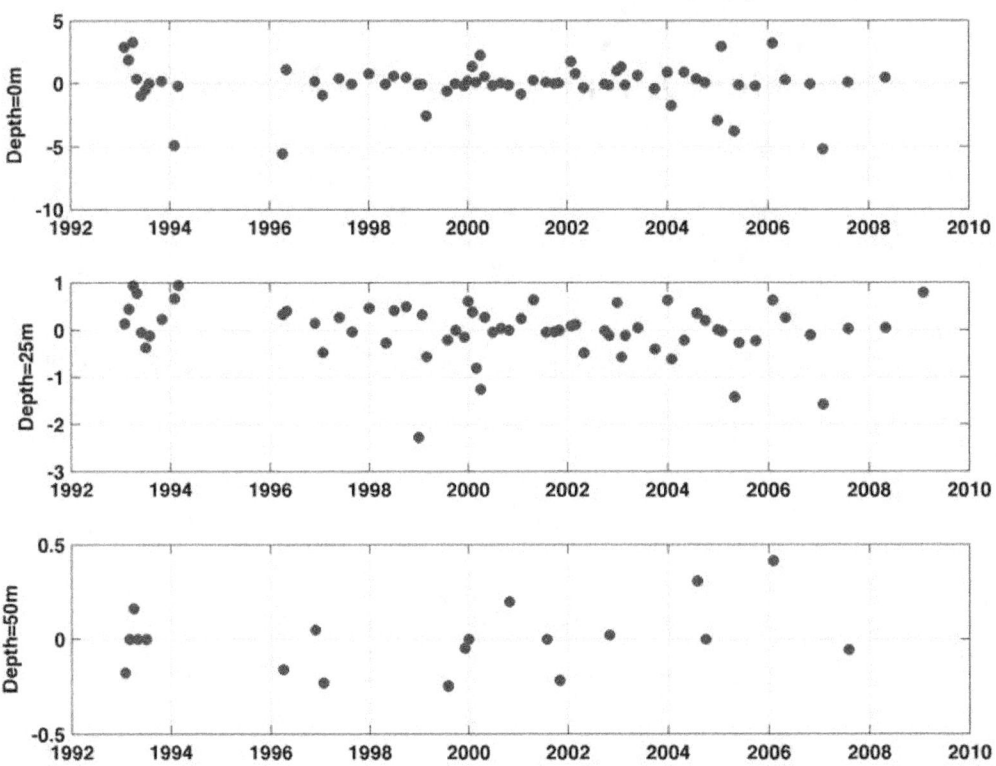

Glacier Bay Oceanographic Monitoring Program Analysis of Observations, 1993–2009
Appendix C. Time series of temperature and salinity anomalies at select depths.

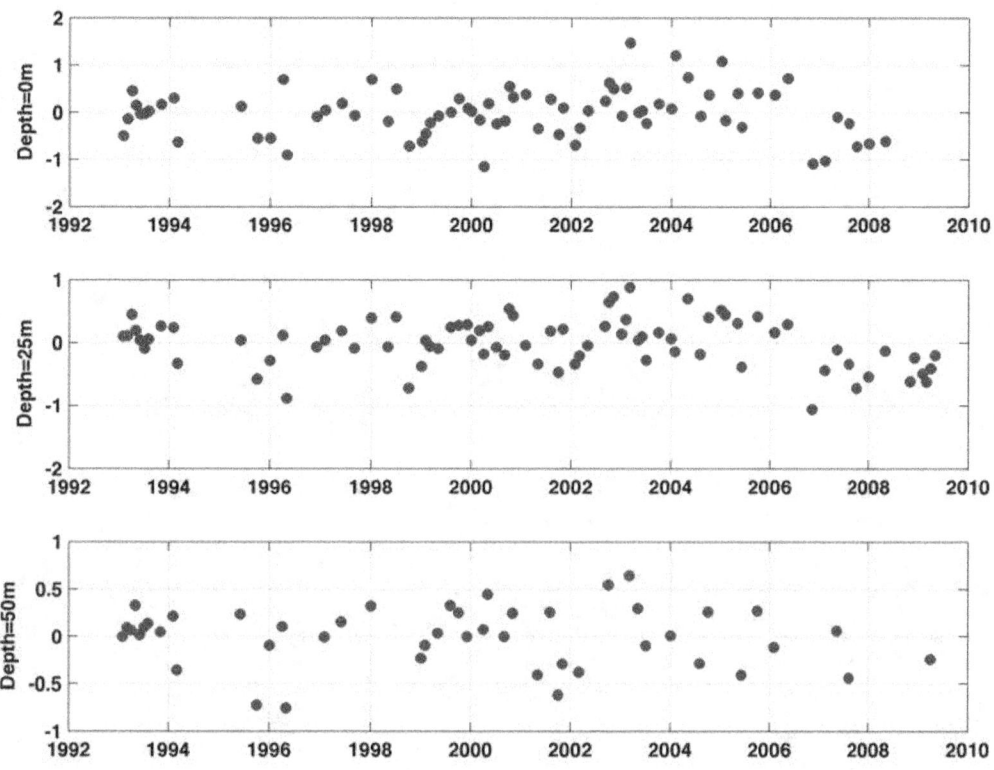

Glacier Bay Oceanographic Monitoring Program Analysis of Observations, 1993–2009
Appendix C. Time series of temperature and salinity anomalies at select depths.

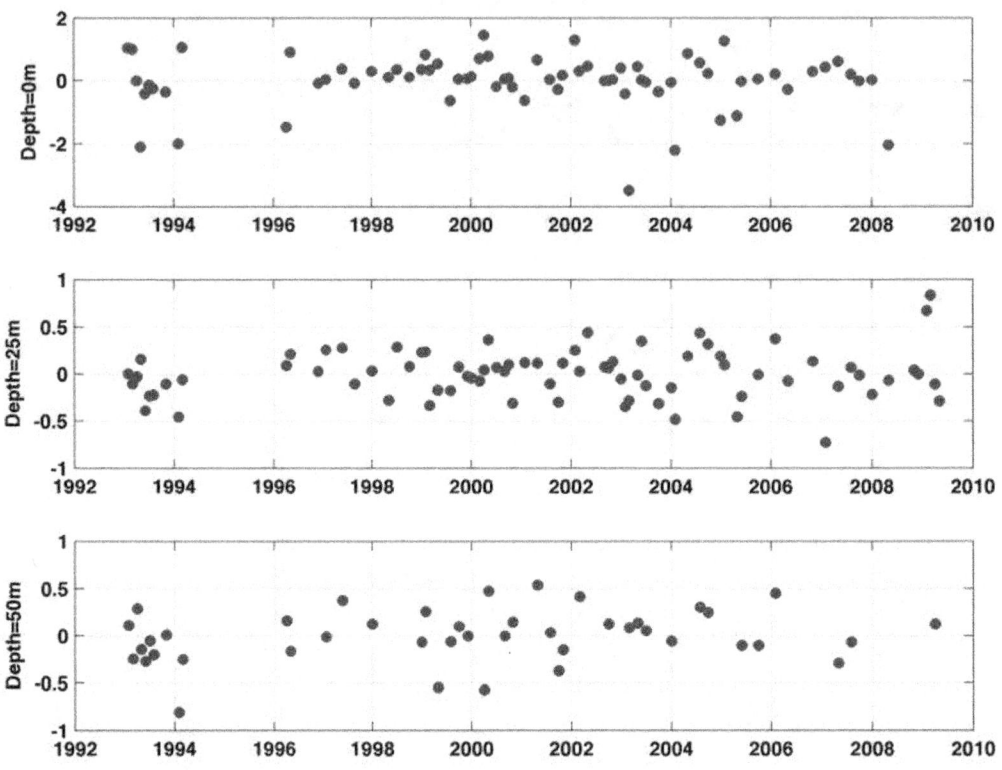

Glacier Bay Oceanographic Monitoring Program Analysis of Observations, 1993–2009
Appendix C. Time series of temperature and salinity anomalies at select depths.

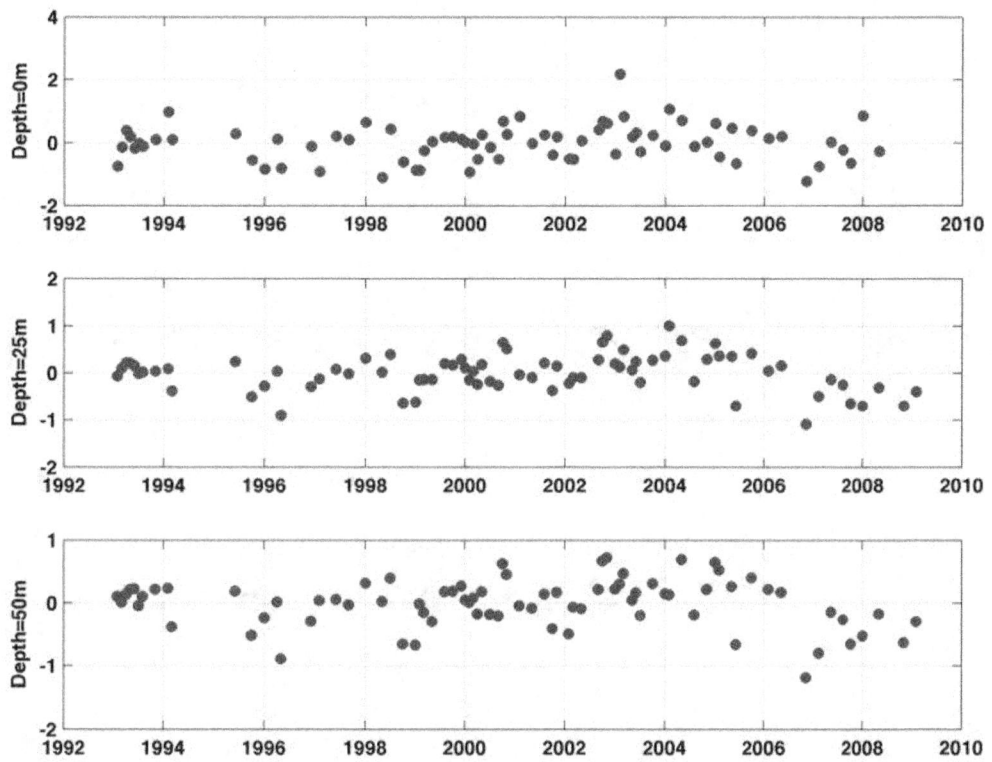

Glacier Bay Oceanographic Monitoring Program Analysis of Observations, 1993–2009
Appendix C. Time series of temperature and salinity anomalies at select depths.

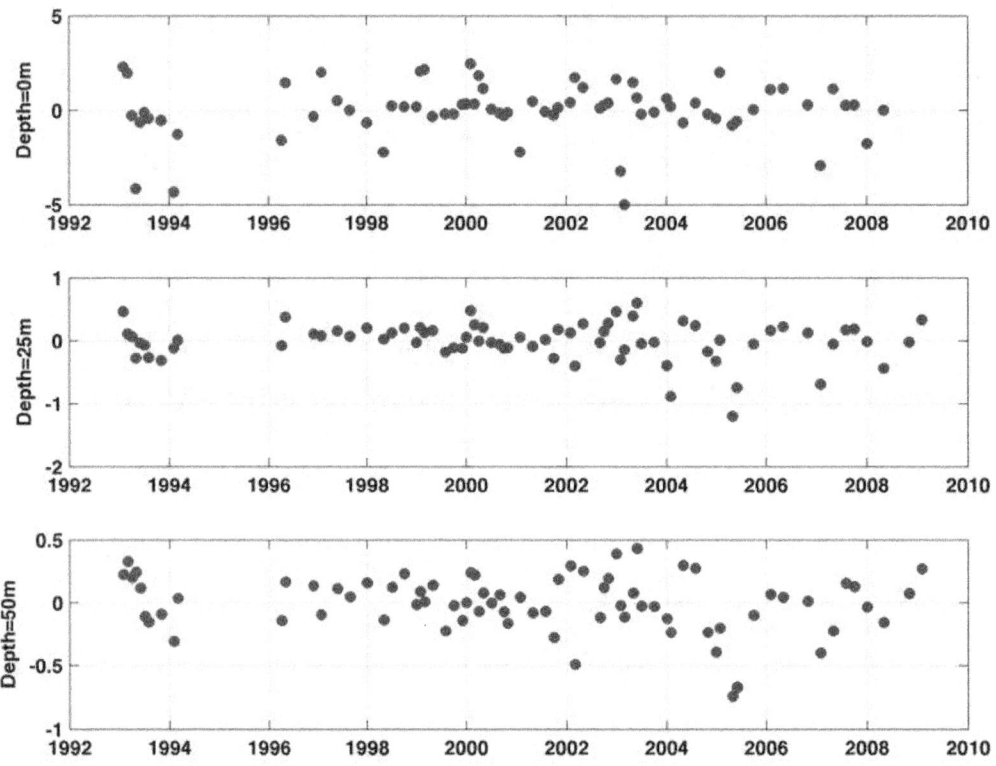

Glacier Bay Oceanographic Monitoring Program Analysis of Observations, 1993–2009
Appendix C. Time series of temperature and salinity anomalies at select depths.

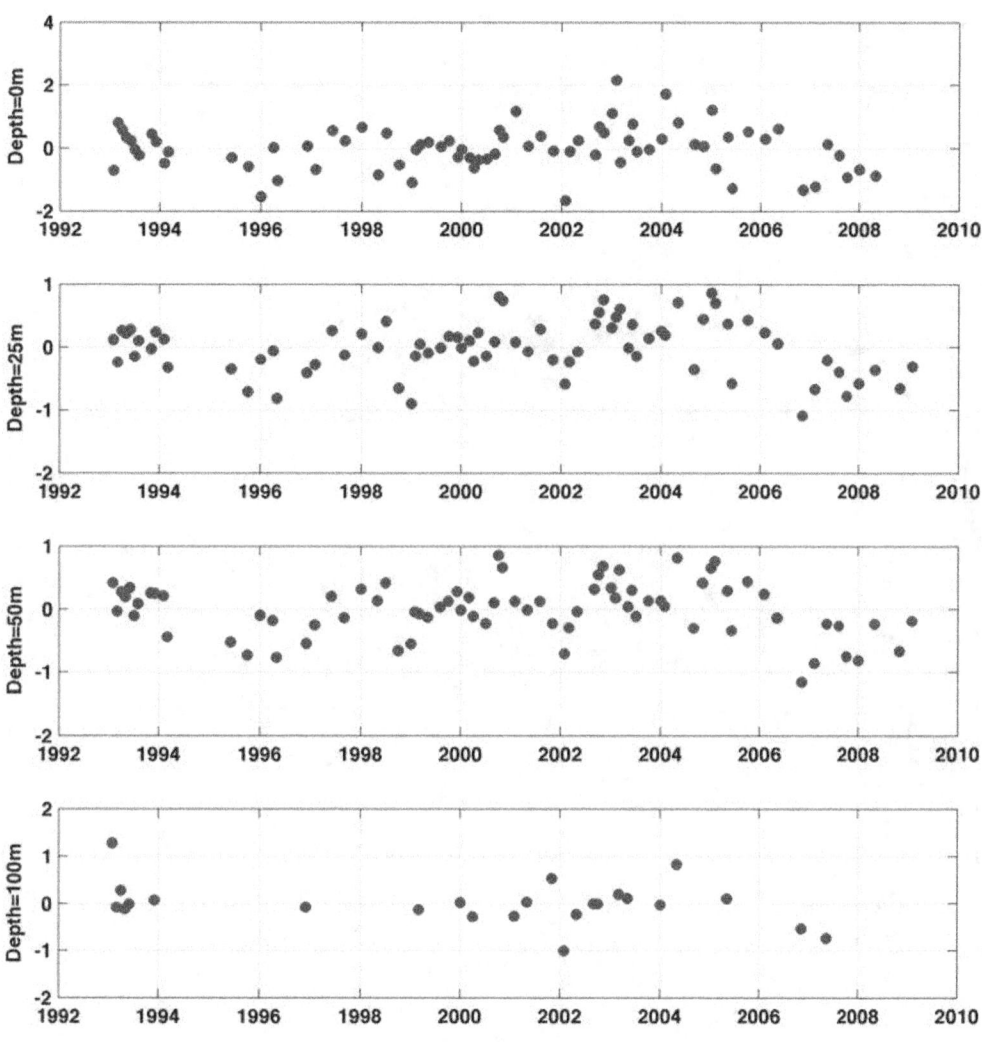

Glacier Bay Oceanographic Monitoring Program Analysis of Observations, 1993–2009
Appendix C. Time series of temperature and salinity anomalies at select depths.

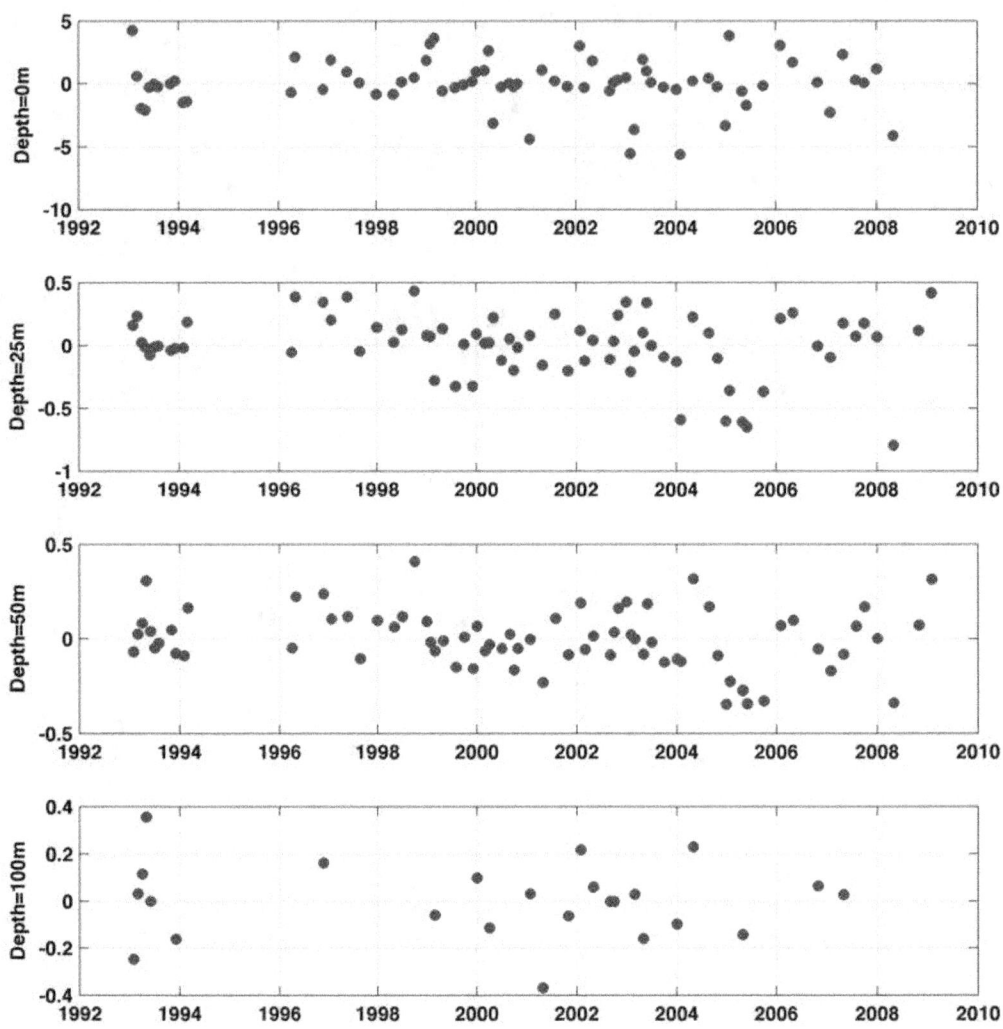

Glacier Bay Oceanographic Monitoring Program Analysis of Observations, 1993–2009
Appendix C. Time series of temperature and salinity anomalies at select depths.

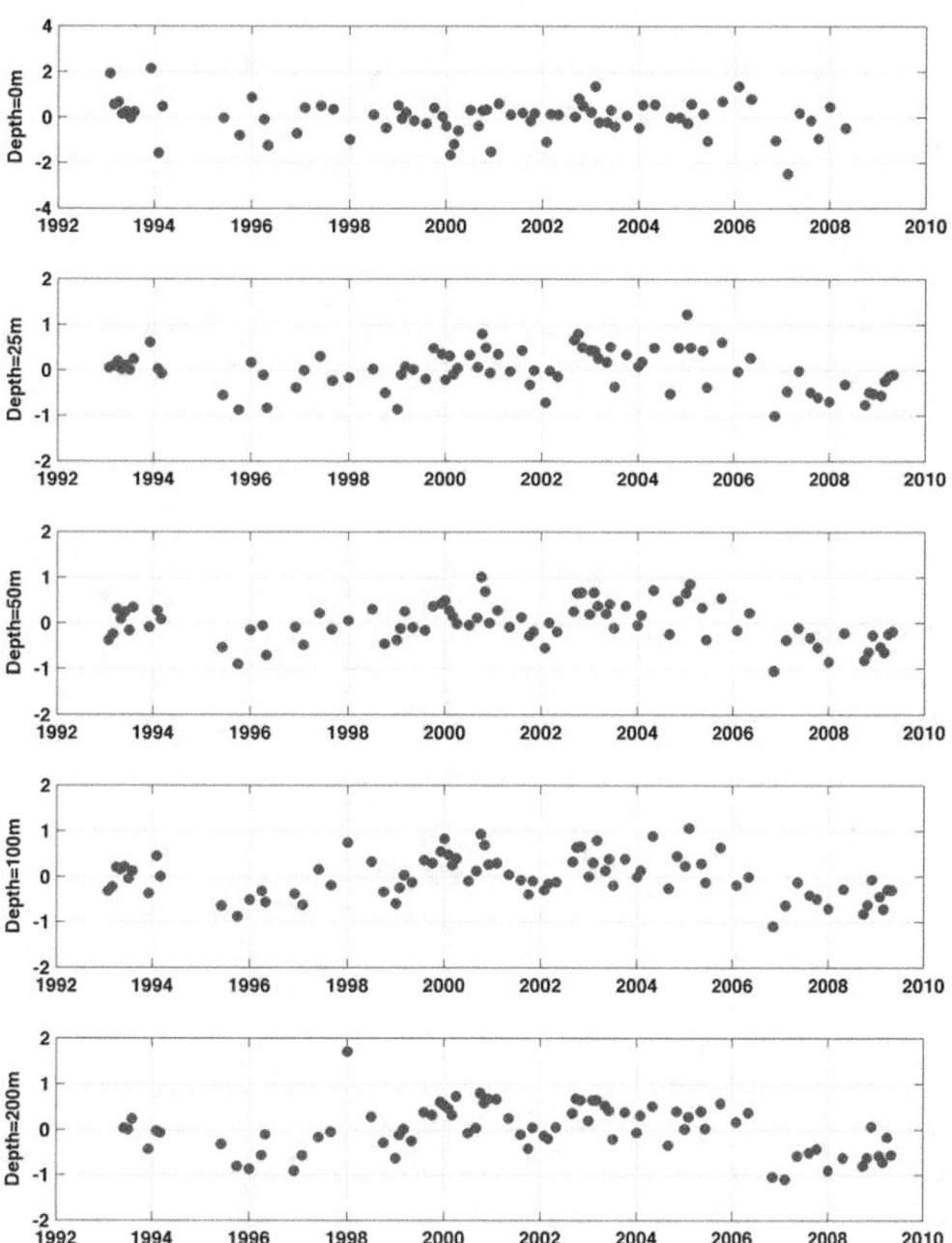

Glacier Bay Oceanographic Monitoring Program Analysis of Observations, 1993–2009
Appendix C. Time series of temperature and salinity anomalies at select depths.

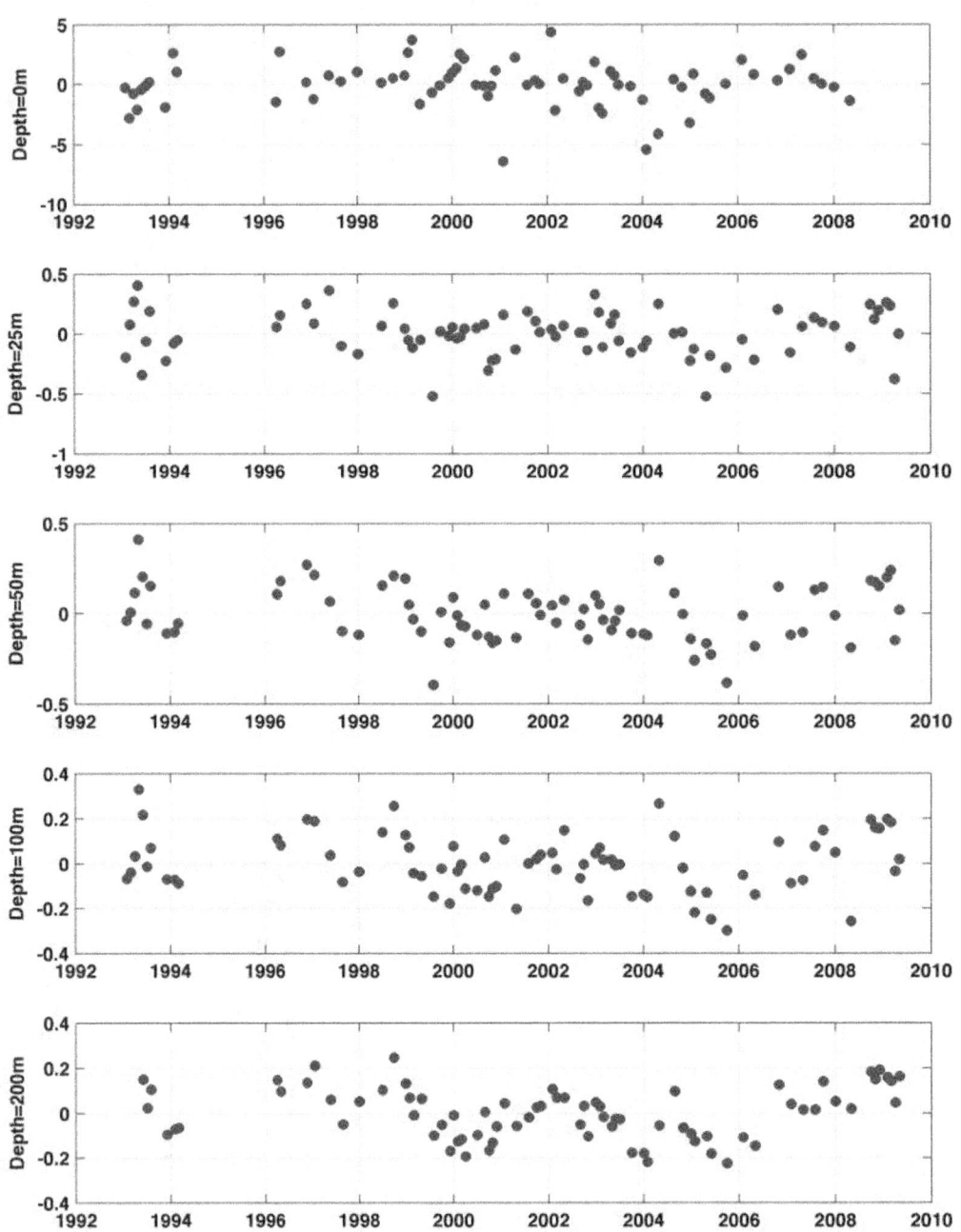

Glacier Bay Oceanographic Monitoring Program Analysis of Observations, 1993–2009
Appendix C. Time series of temperature and salinity anomalies at select depths.

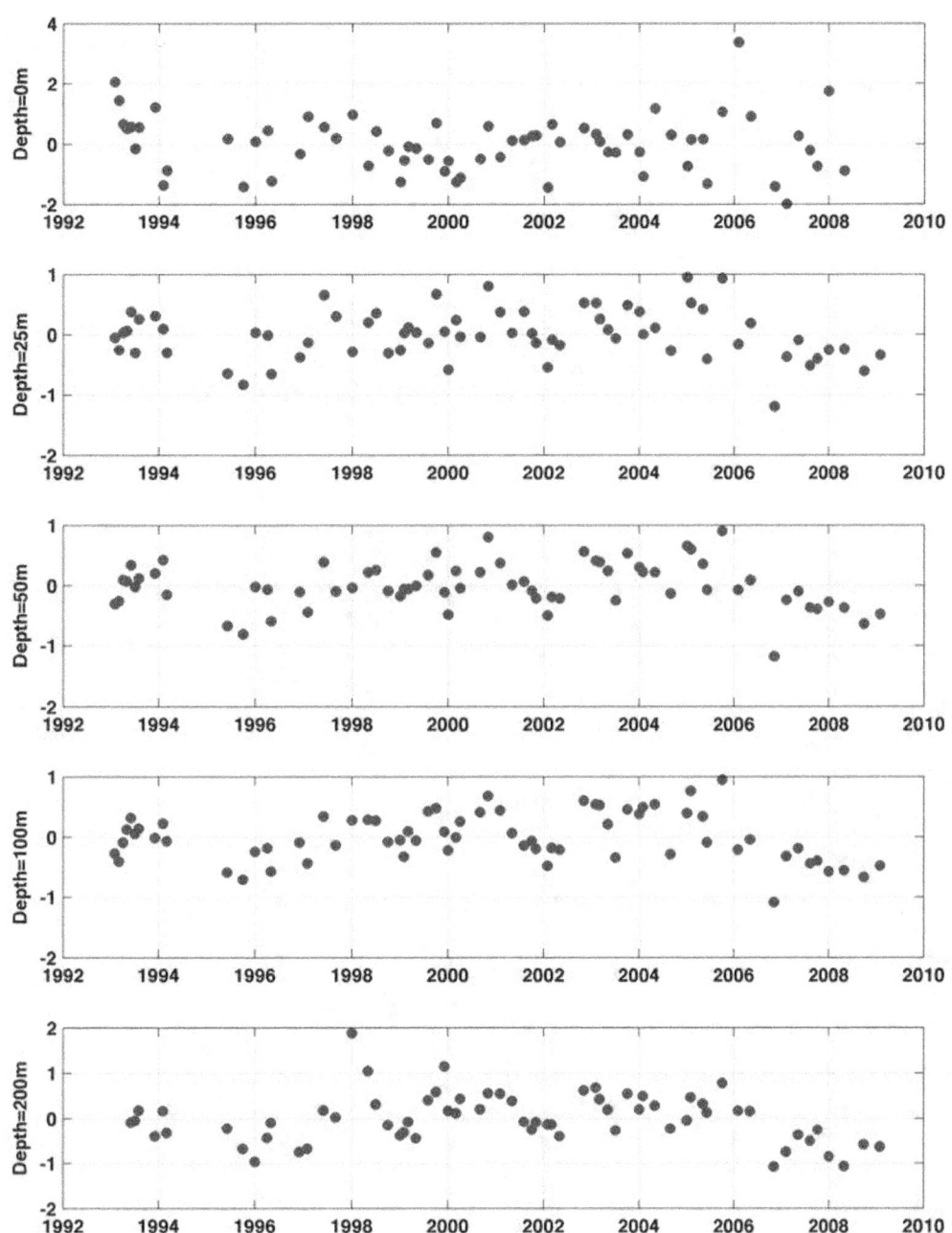

Glacier Bay Oceanographic Monitoring Program Analysis of Observations, 1993–2009
Appendix C. Time series of temperature and salinity anomalies at select depths.

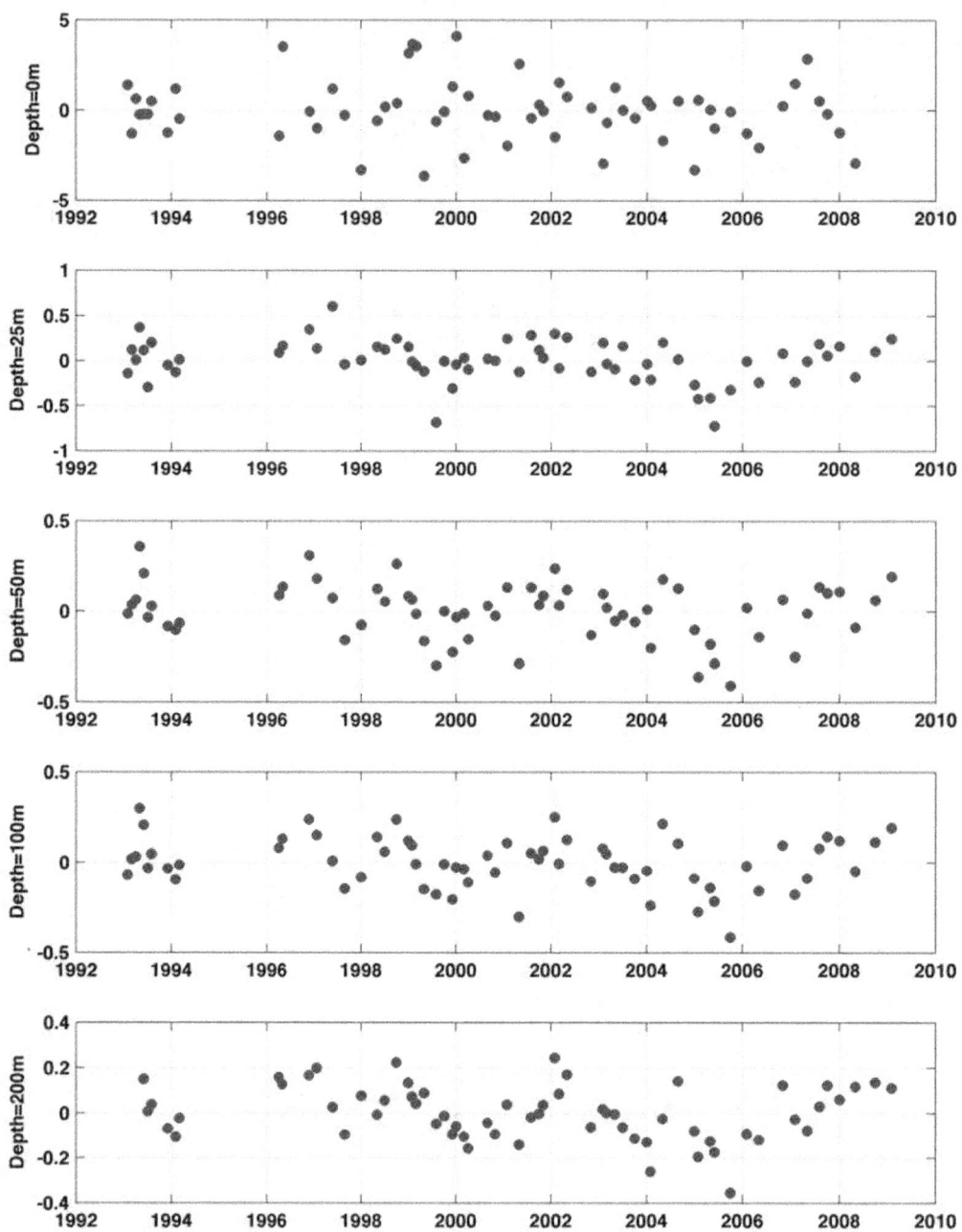

Glacier Bay Oceanographic Monitoring Program Analysis of Observations, 1993–2009
Appendix C. Time series of temperature and salinity anomalies at select depths.

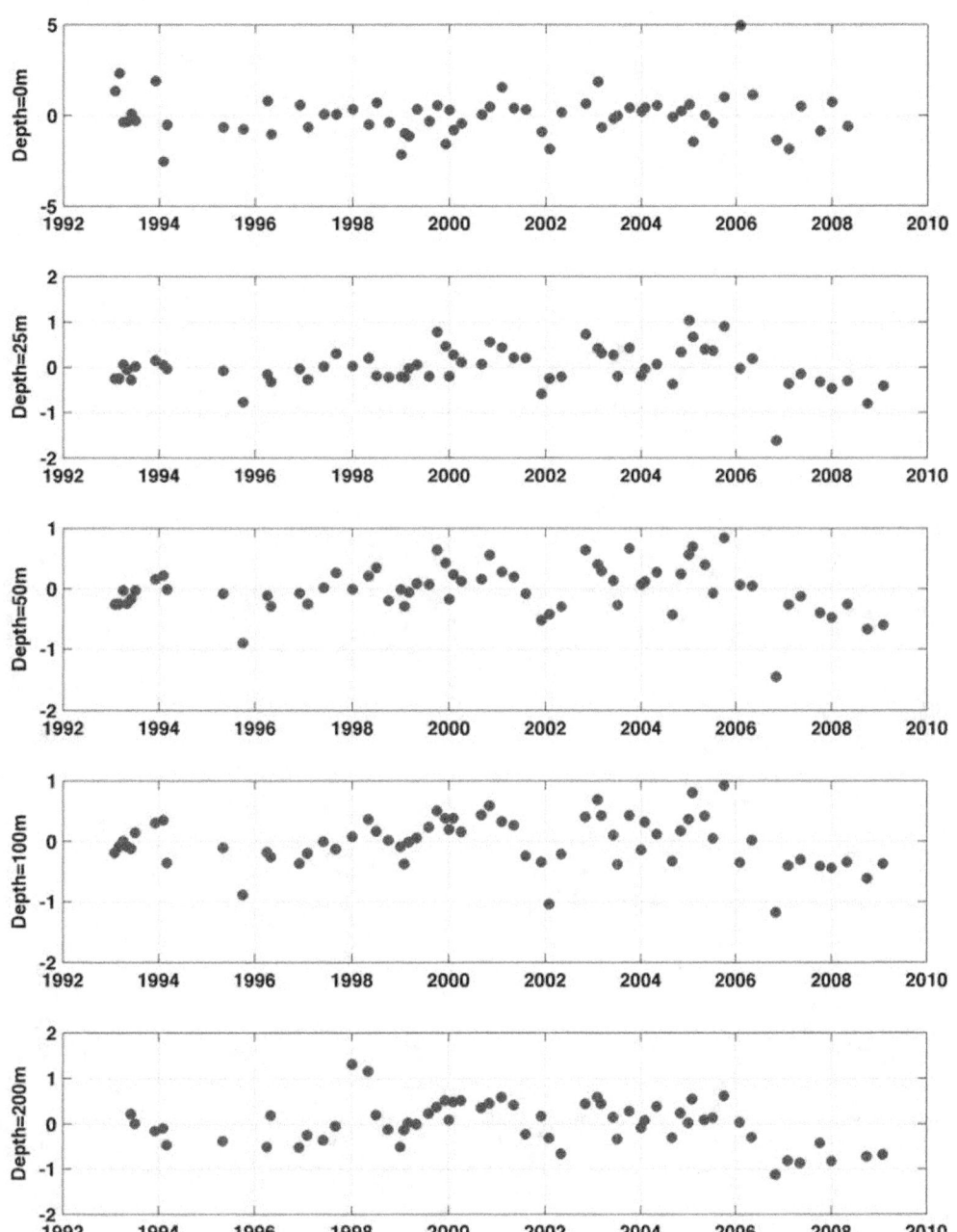

Glacier Bay Oceanographic Monitoring Program Analysis of Observations, 1993–2009
Appendix C. Time series of temperature and salinity anomalies at select depths.

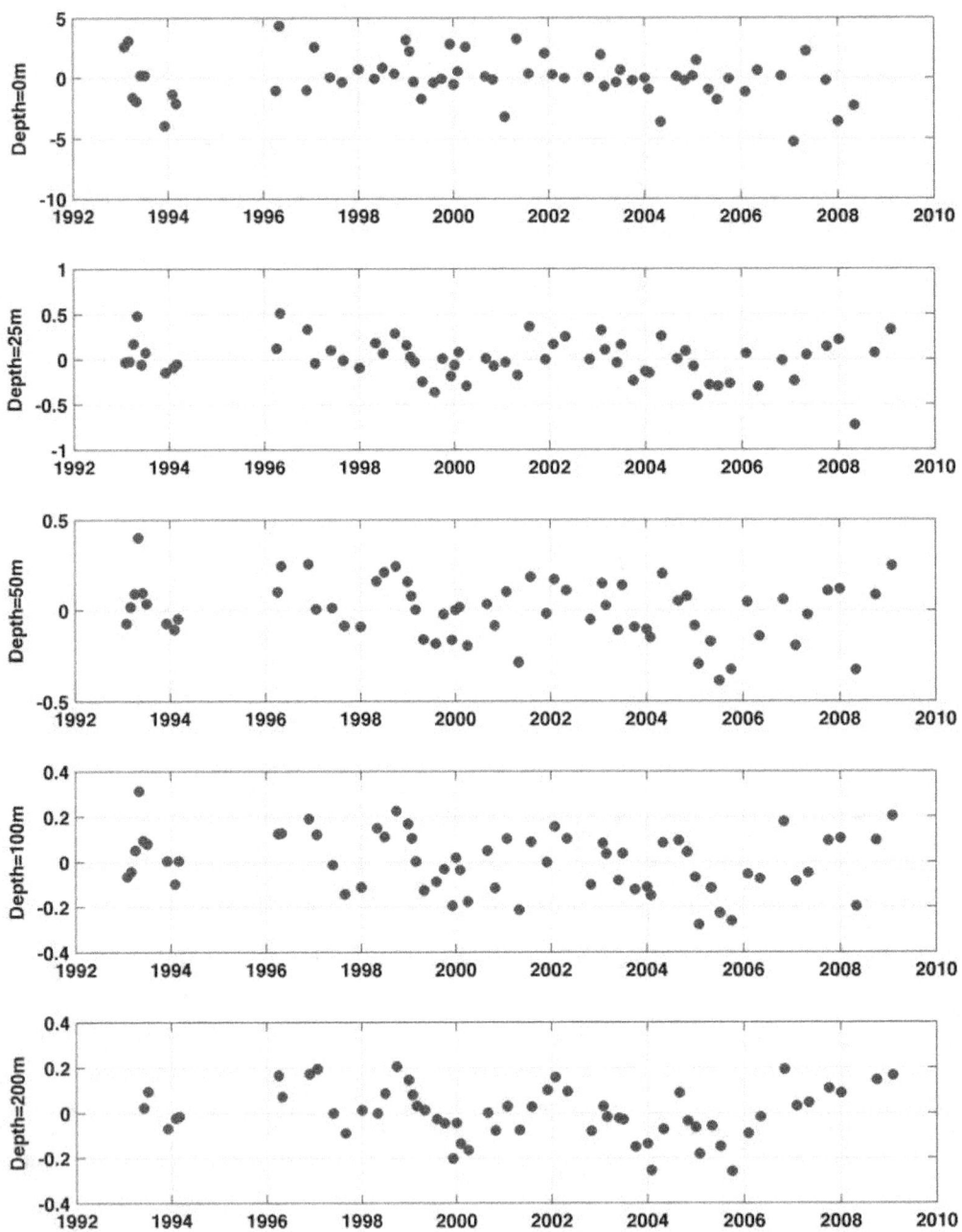

Glacier Bay Oceanographic Monitoring Program Analysis of Observations, 1993–2009
Appendix C. Time series of temperature and salinity anomalies at select depths.

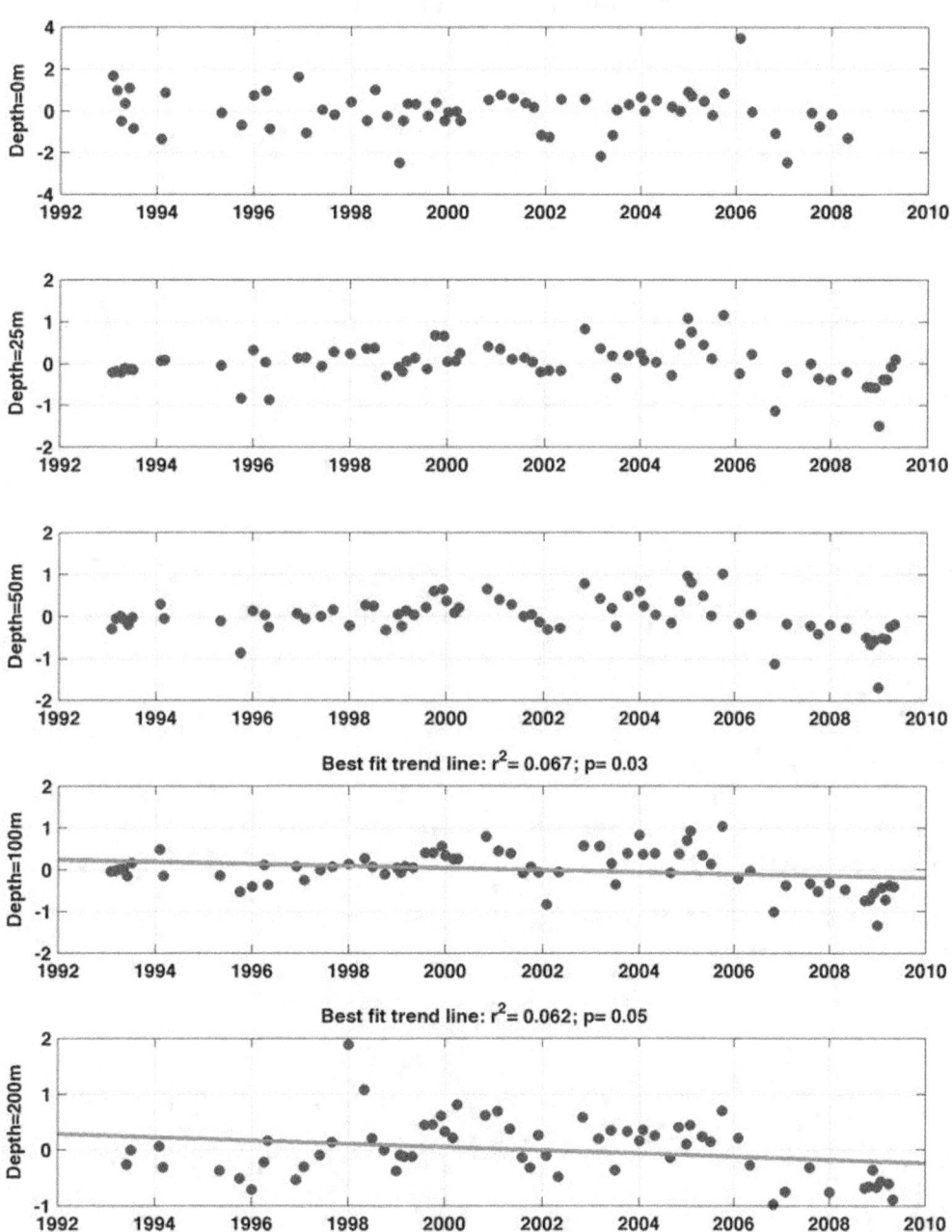

Glacier Bay Oceanographic Monitoring Program Analysis of Observations, 1993–2009
Appendix C. Time series of temperature and salinity anomalies at select depths.

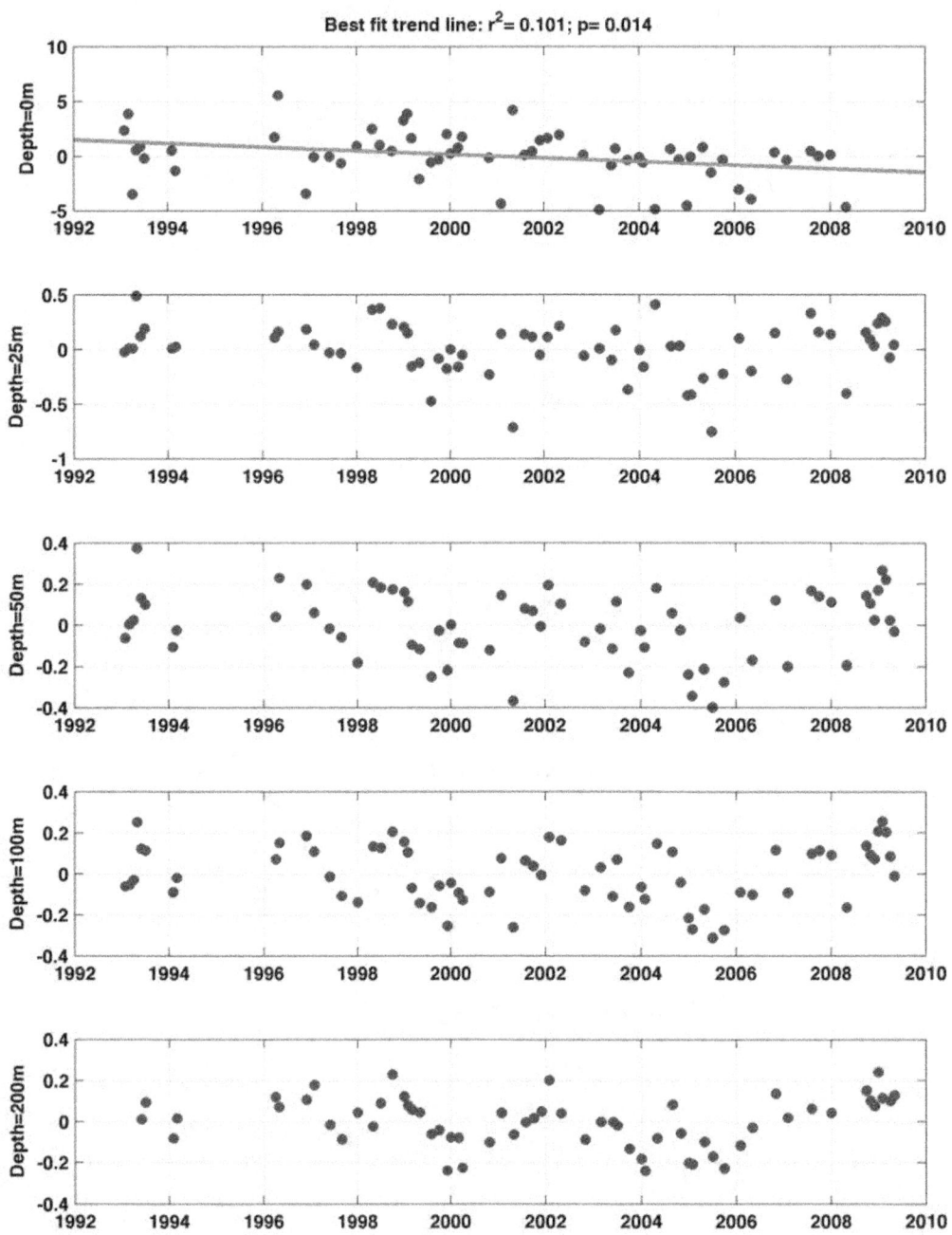

Glacier Bay Oceanographic Monitoring Program Analysis of Observations, 1993–2009
Appendix C. Time series of temperature and salinity anomalies at select depths.

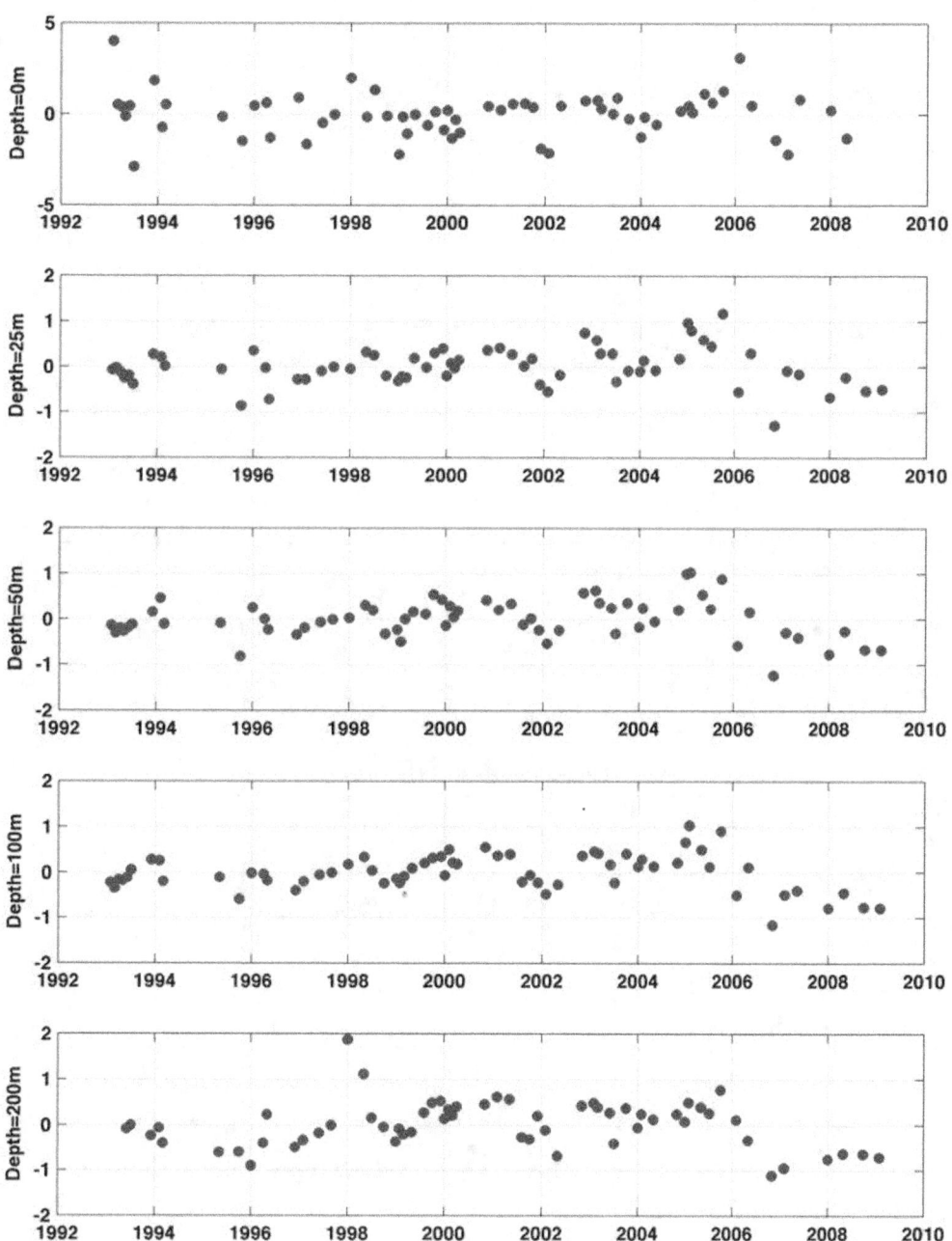

Glacier Bay Oceanographic Monitoring Program Analysis of Observations, 1993–2009
Appendix C. Time series of temperature and salinity anomalies at select depths.

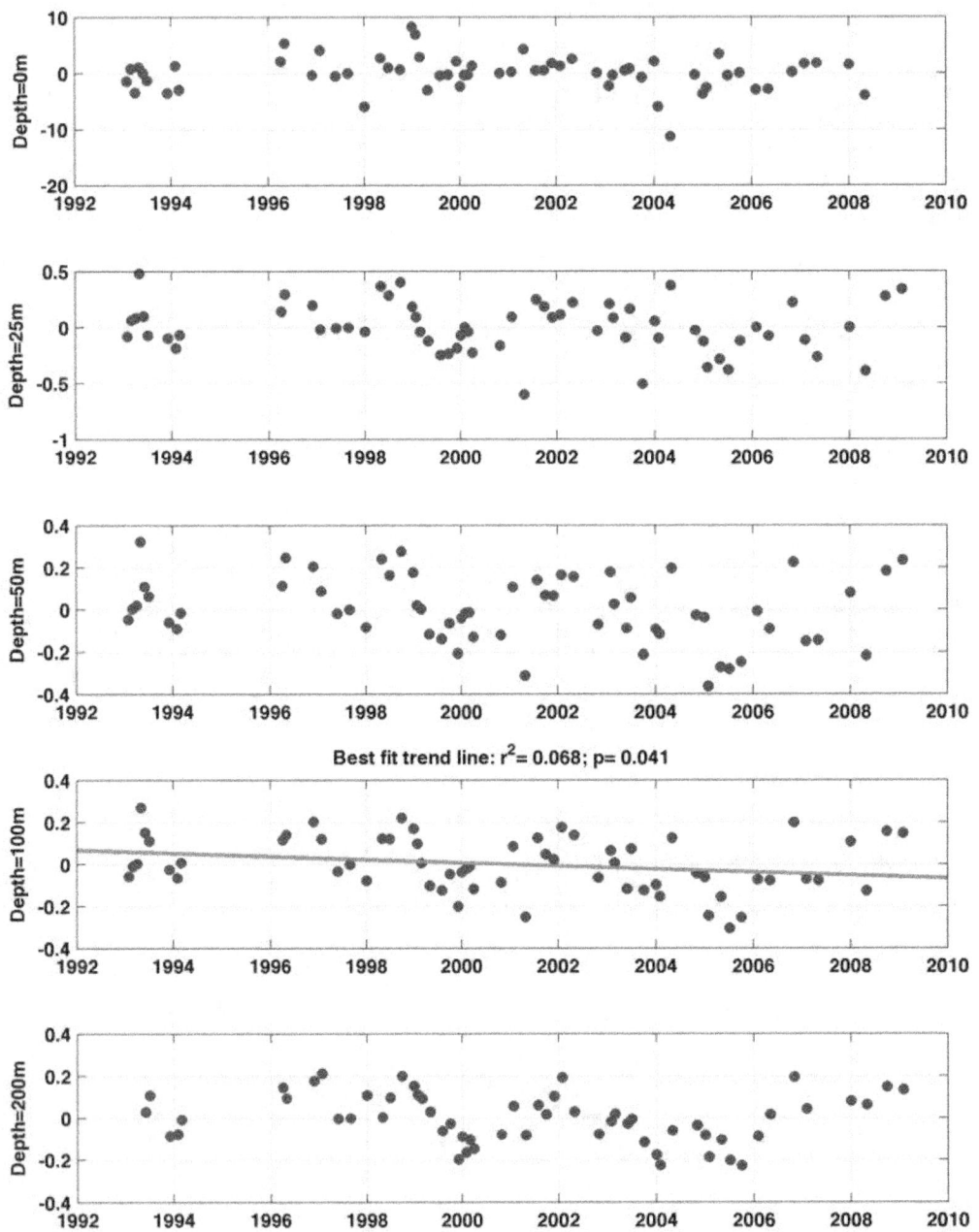

Glacier Bay Oceanographic Monitoring Program Analysis of Observations, 1993–2009
Appendix C. Time series of temperature and salinity anomalies at select depths.

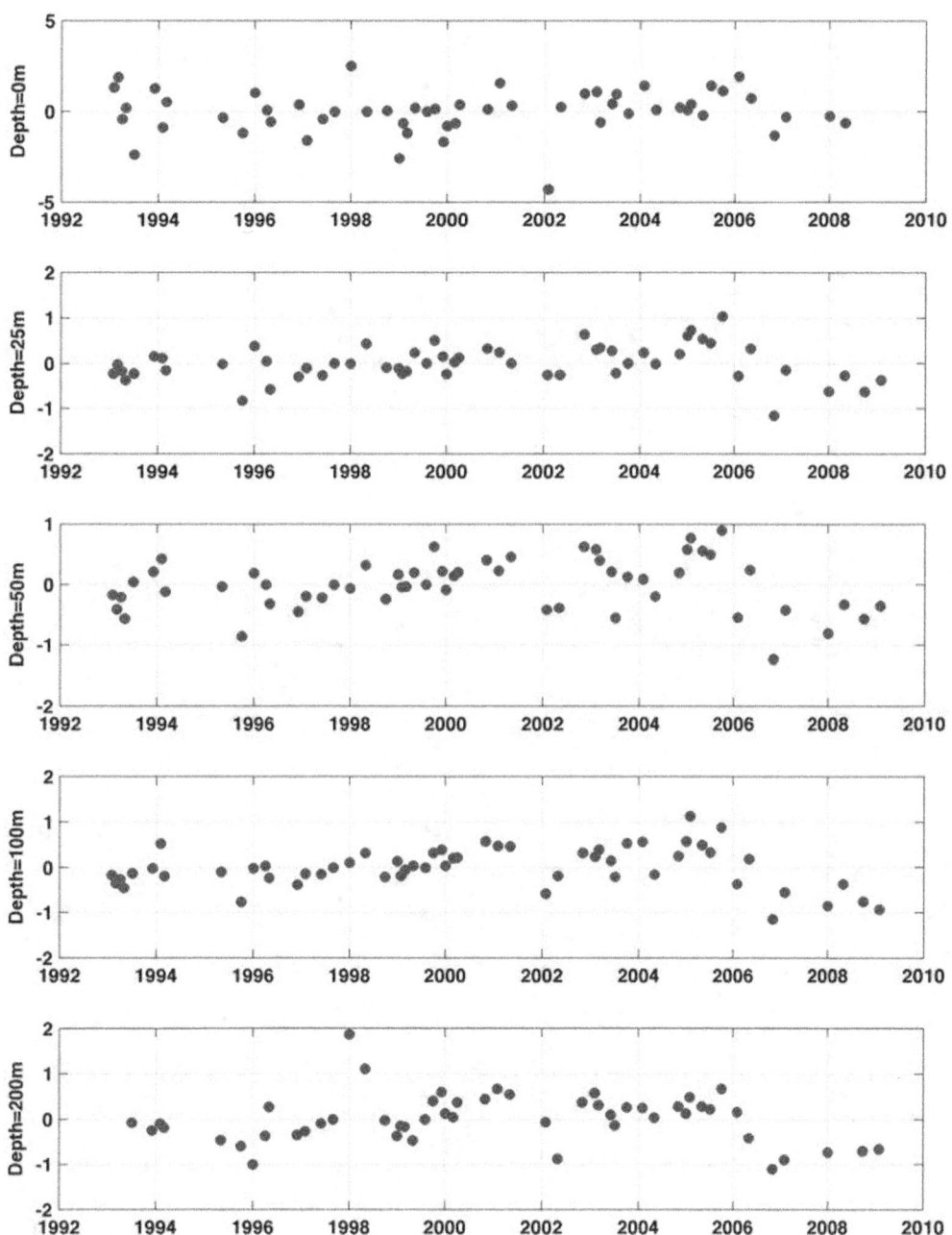

Glacier Bay Oceanographic Monitoring Program Analysis of Observations, 1993–2009
Appendix C. Time series of temperature and salinity anomalies at select depths.

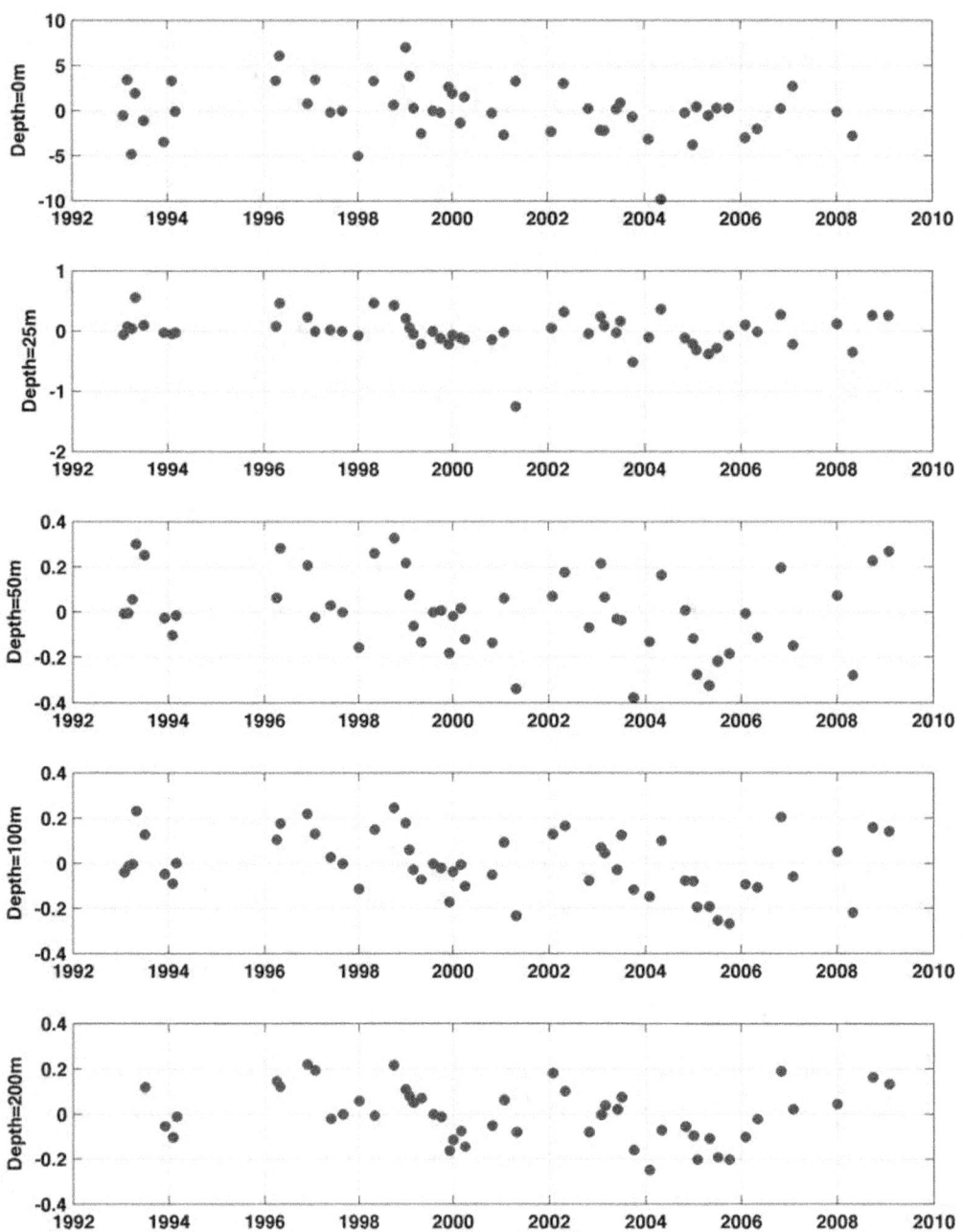

Glacier Bay Oceanographic Monitoring Program Analysis of Observations, 1993–2009
Appendix C. Time series of temperature and salinity anomalies at select depths.

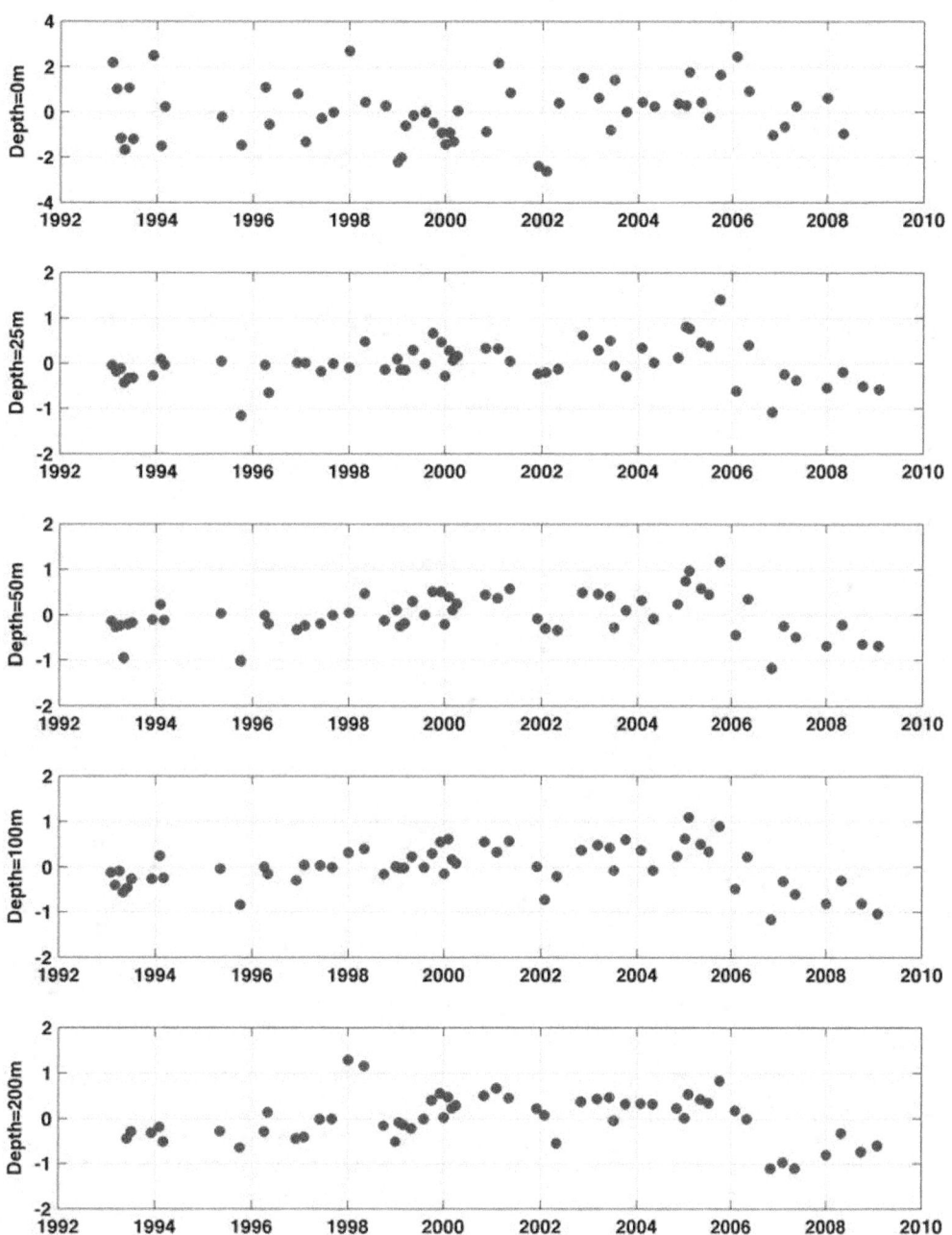

Glacier Bay Oceanographic Monitoring Program Analysis of Observations, 1993–2009
Appendix C. Time series of temperature and salinity anomalies at select depths.

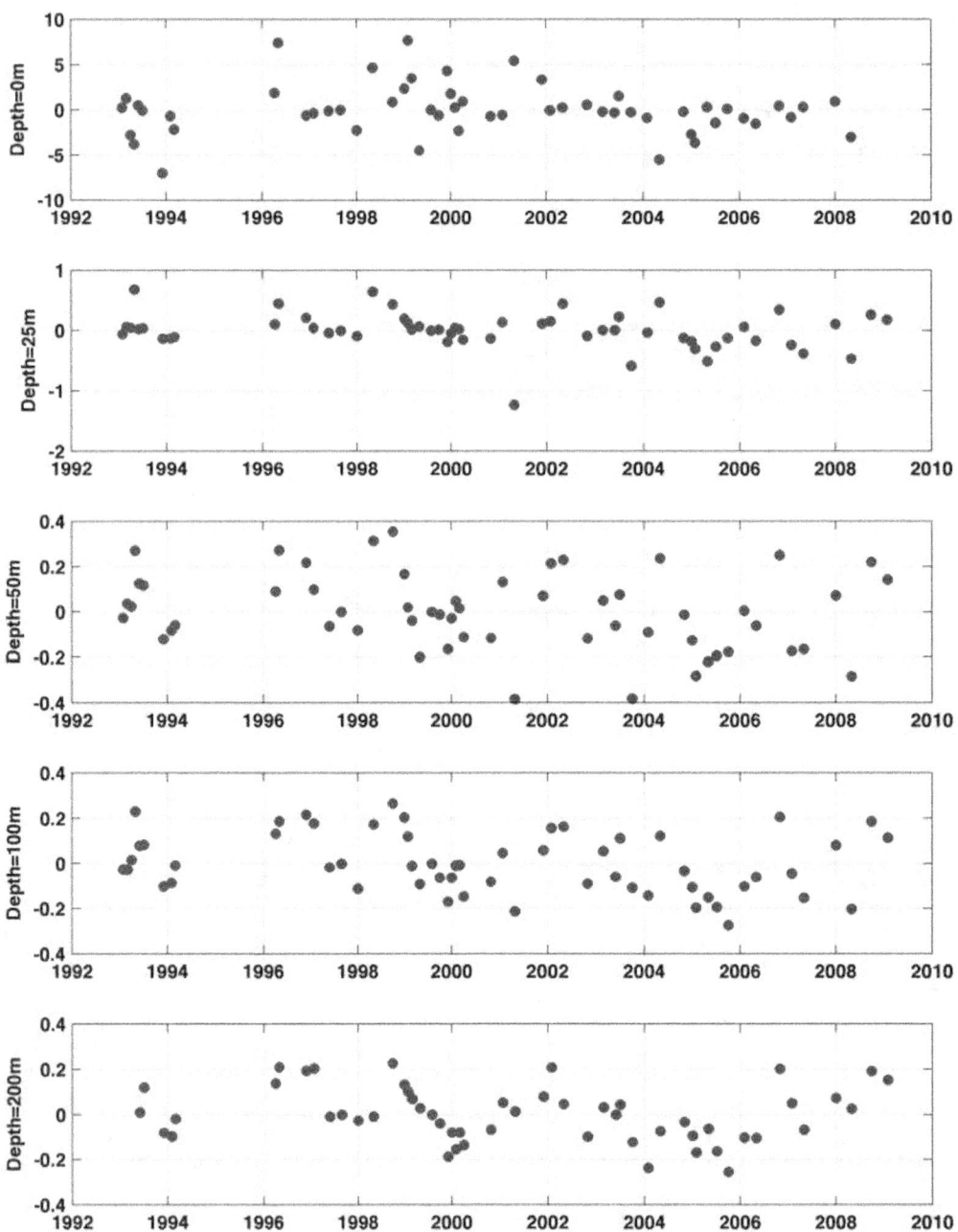

Glacier Bay Oceanographic Monitoring Program Analysis of Observations, 1993–2009
Appendix C. Time series of temperature and salinity anomalies at select depths.

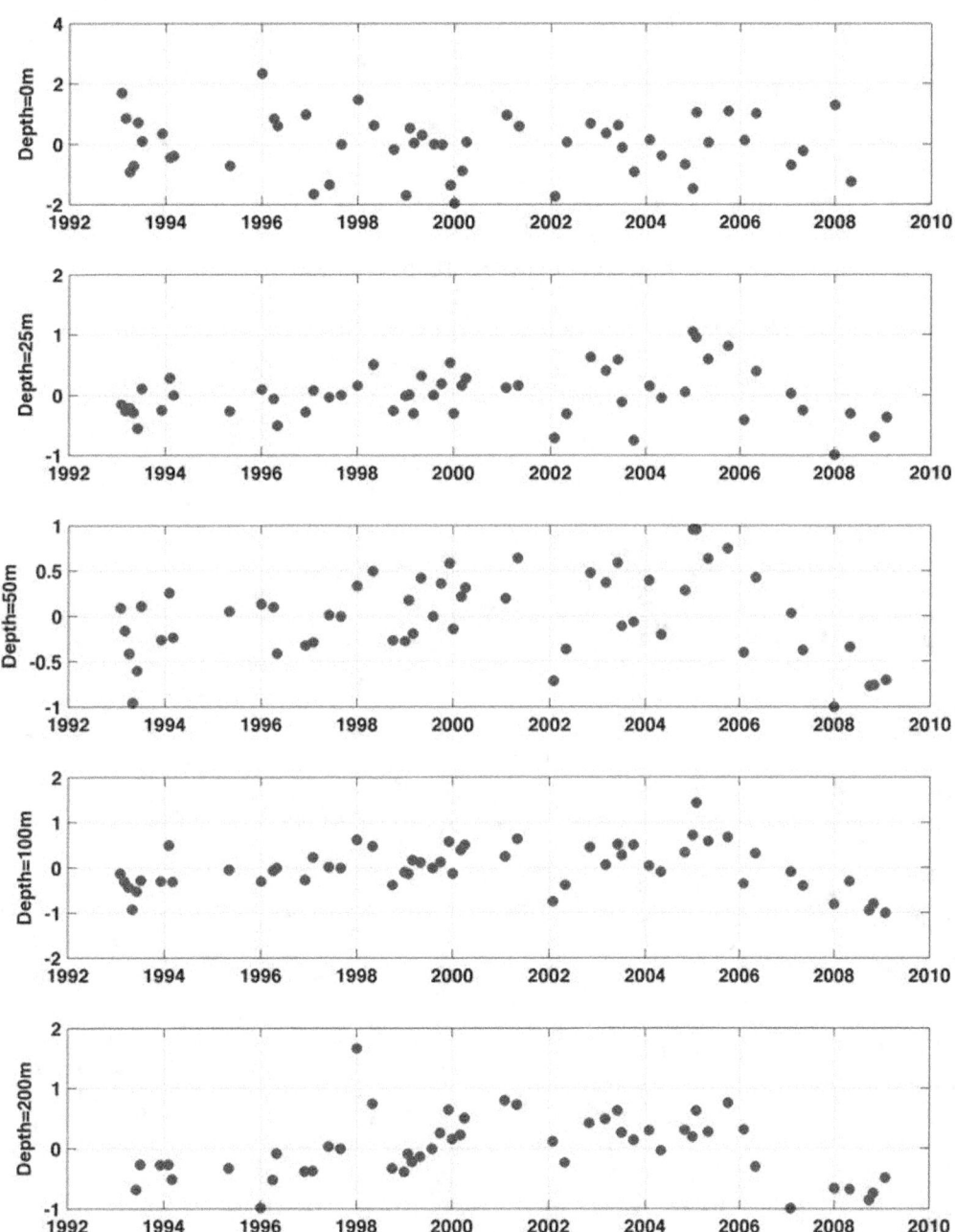

Glacier Bay Oceanographic Monitoring Program Analysis of Observations, 1993–2009
Appendix C. Time series of temperature and salinity anomalies at select depths.

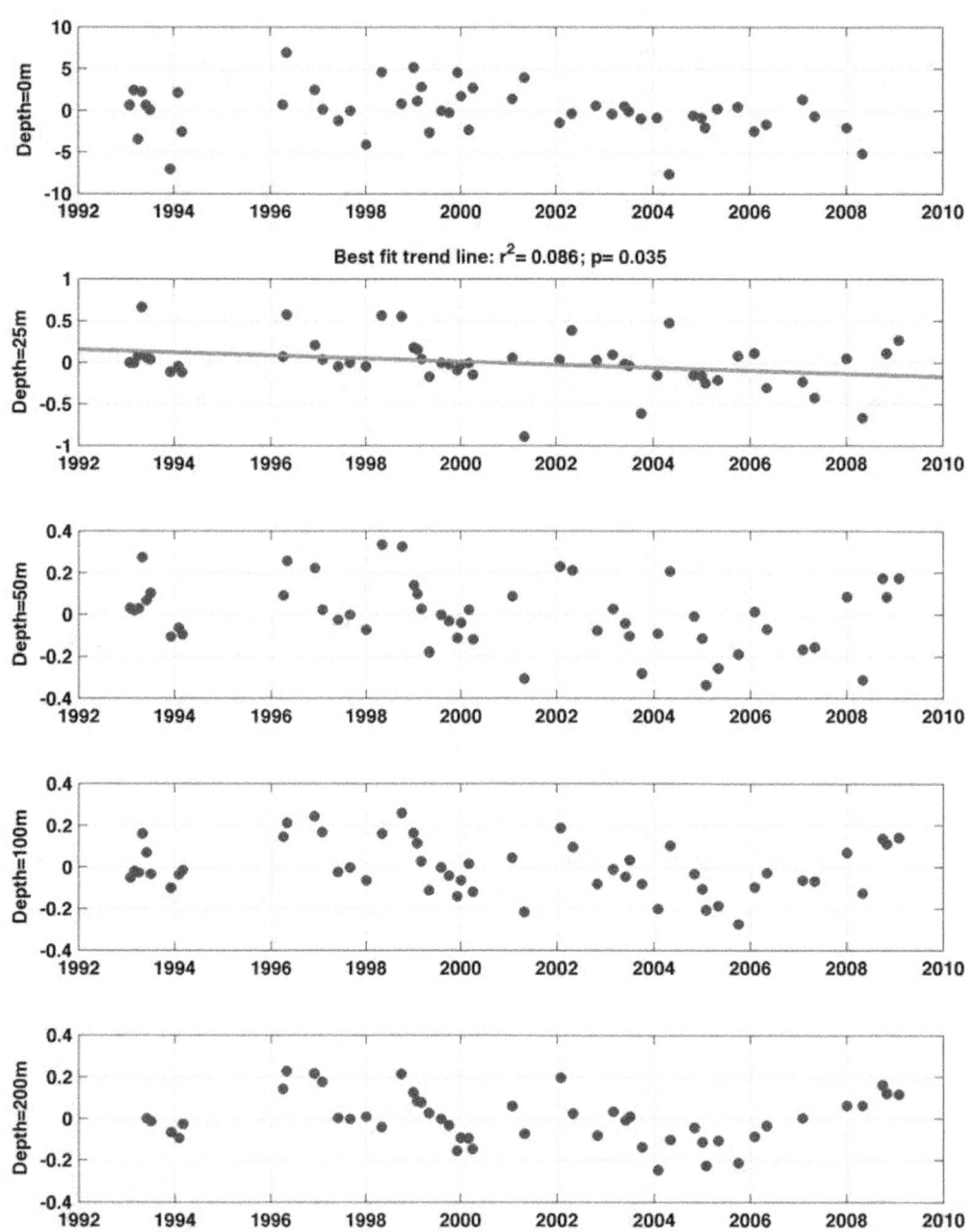

Glacier Bay Oceanographic Monitoring Program Analysis of Observations, 1993–2009
Appendix C. Time series of temperature and salinity anomalies at select depths.

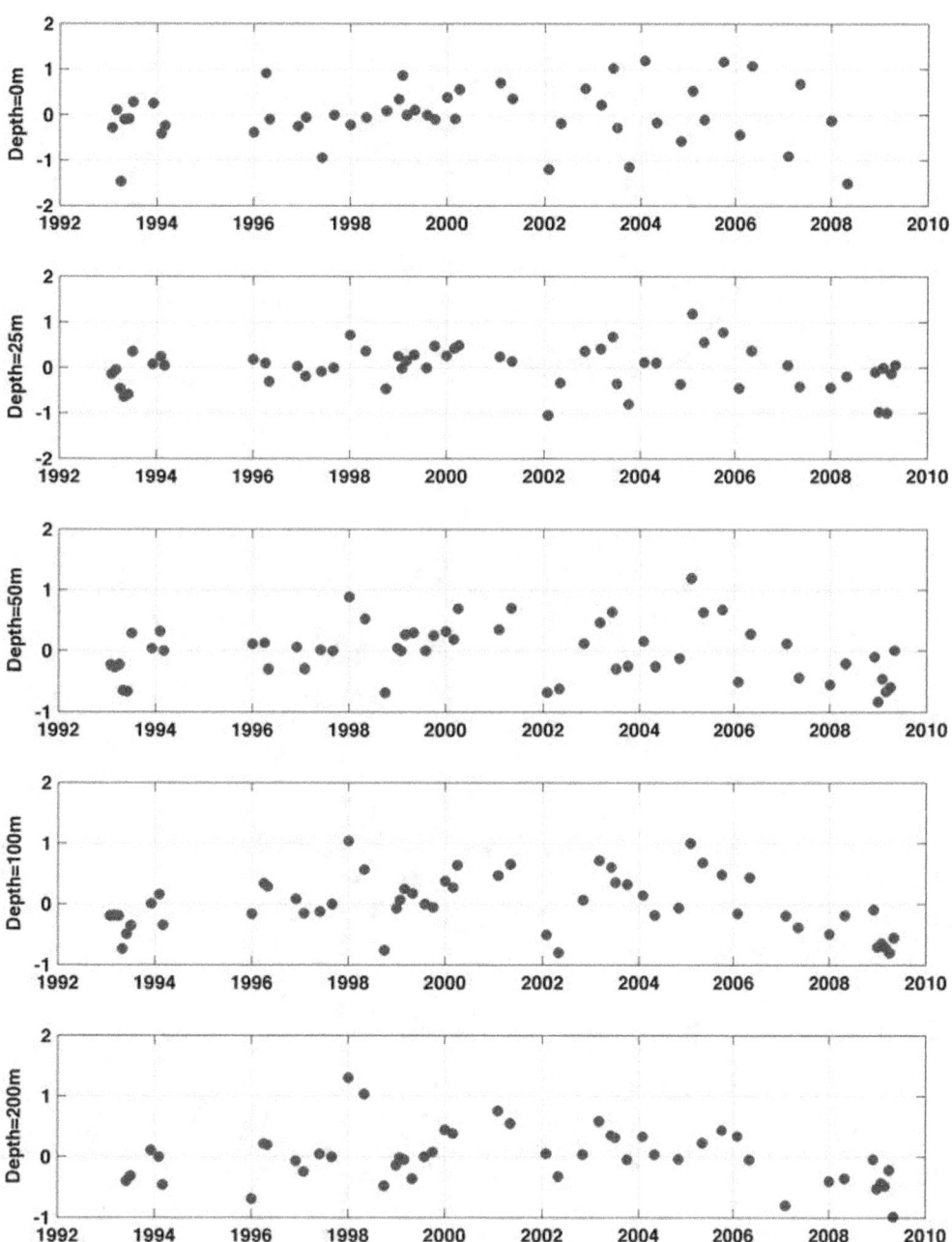

Glacier Bay Oceanographic Monitoring Program Analysis of Observations, 1993–2009
Appendix C. Time series of temperature and salinity anomalies at select depths.

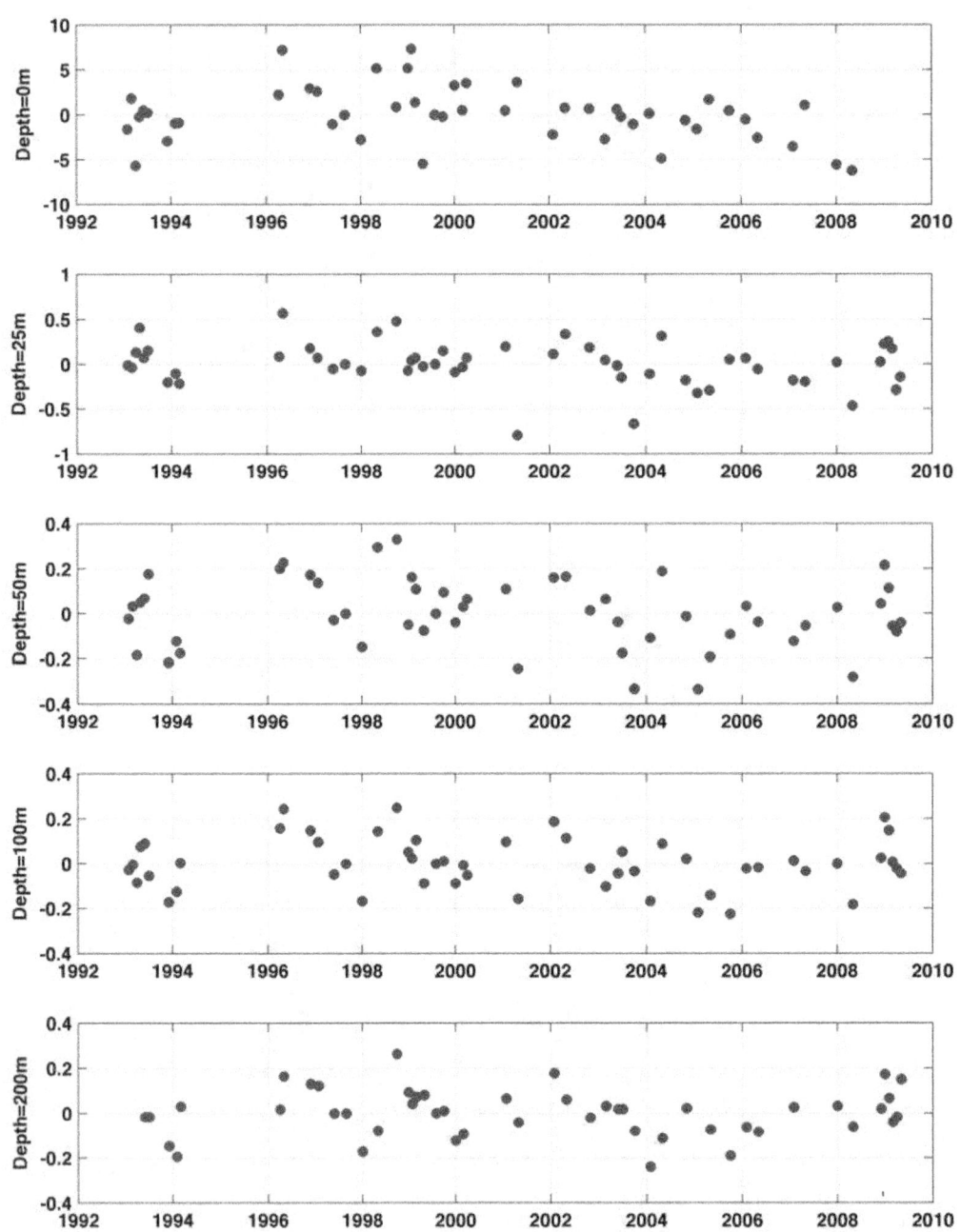

Glacier Bay Oceanographic Monitoring Program Analysis of Observations, 1993–2009
Appendix C. Time series of temperature and salinity anomalies at select depths.

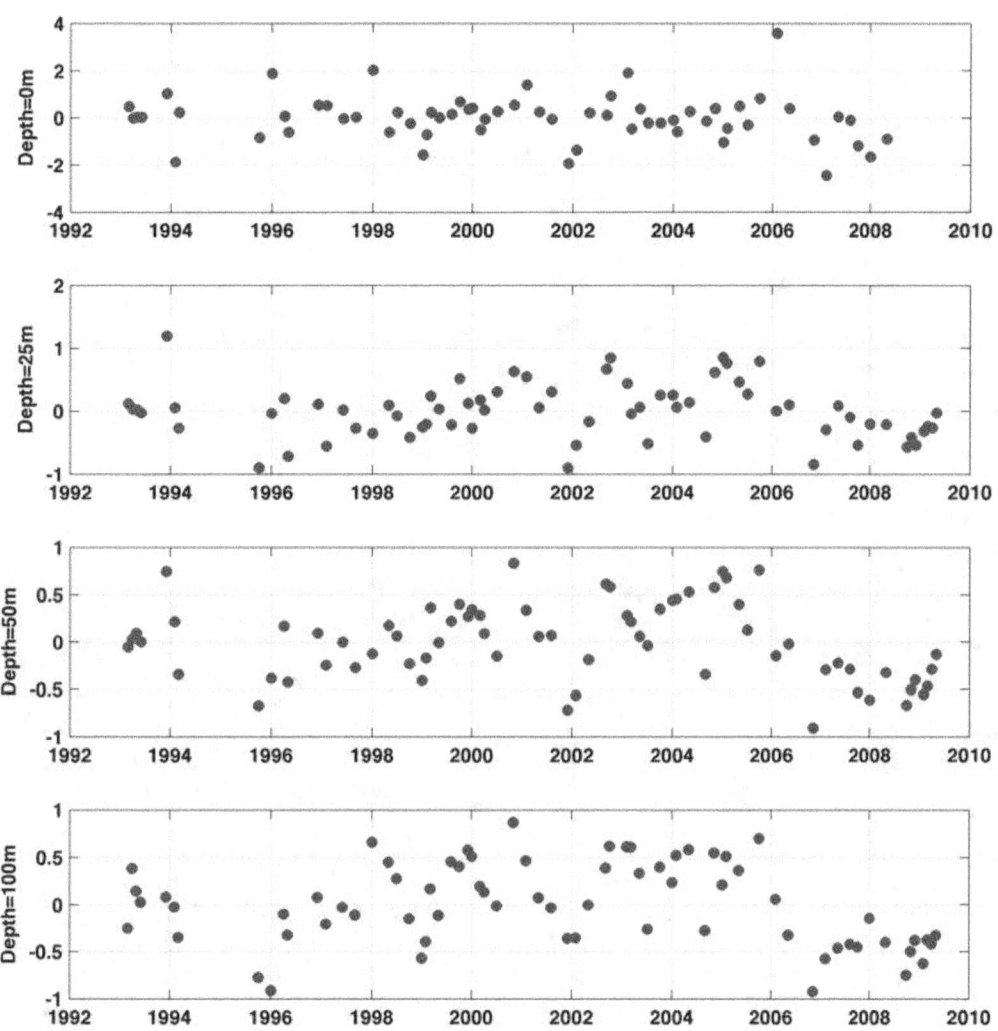

Glacier Bay Oceanographic Monitoring Program Analysis of Observations, 1993–2009
Appendix C. Time series of temperature and salinity anomalies at select depths.

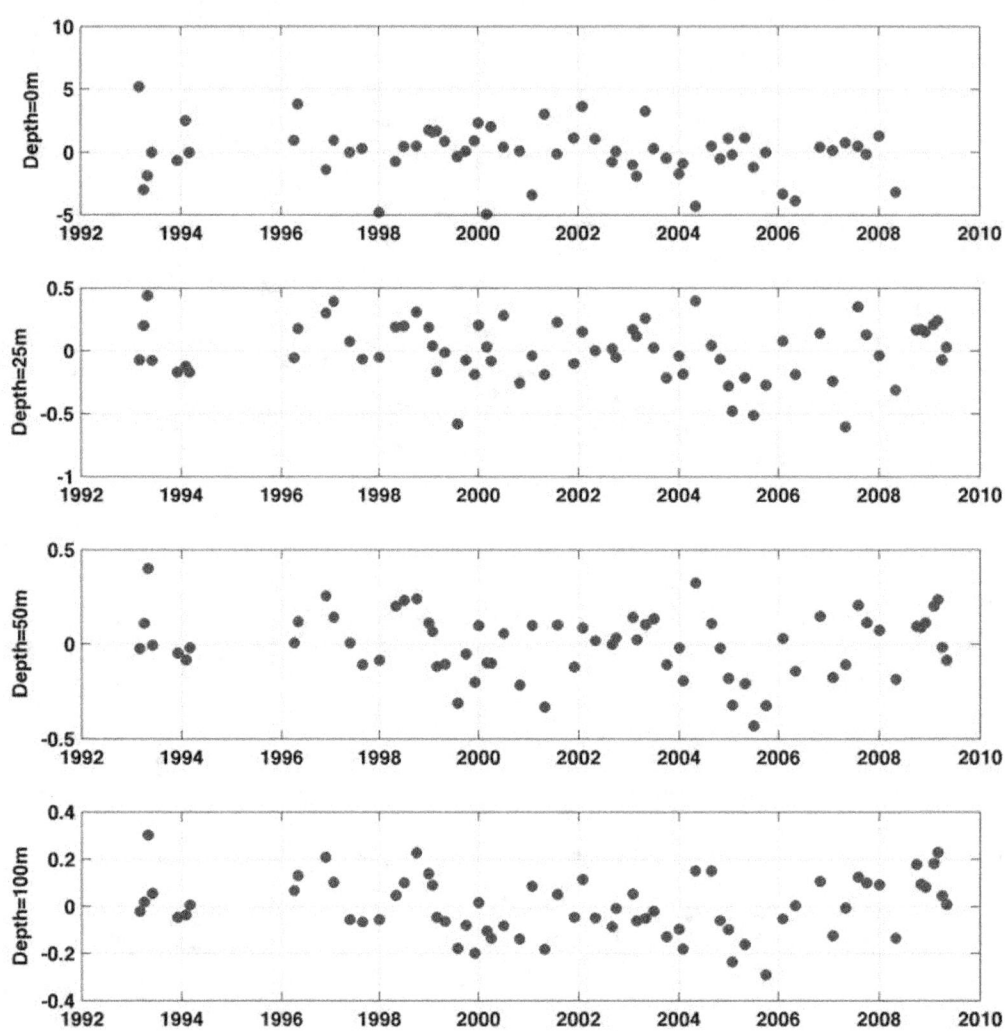

Glacier Bay Oceanographic Monitoring Program Analysis of Observations, 1993–2009
Appendix C. Time series of temperature and salinity anomalies at select depths.

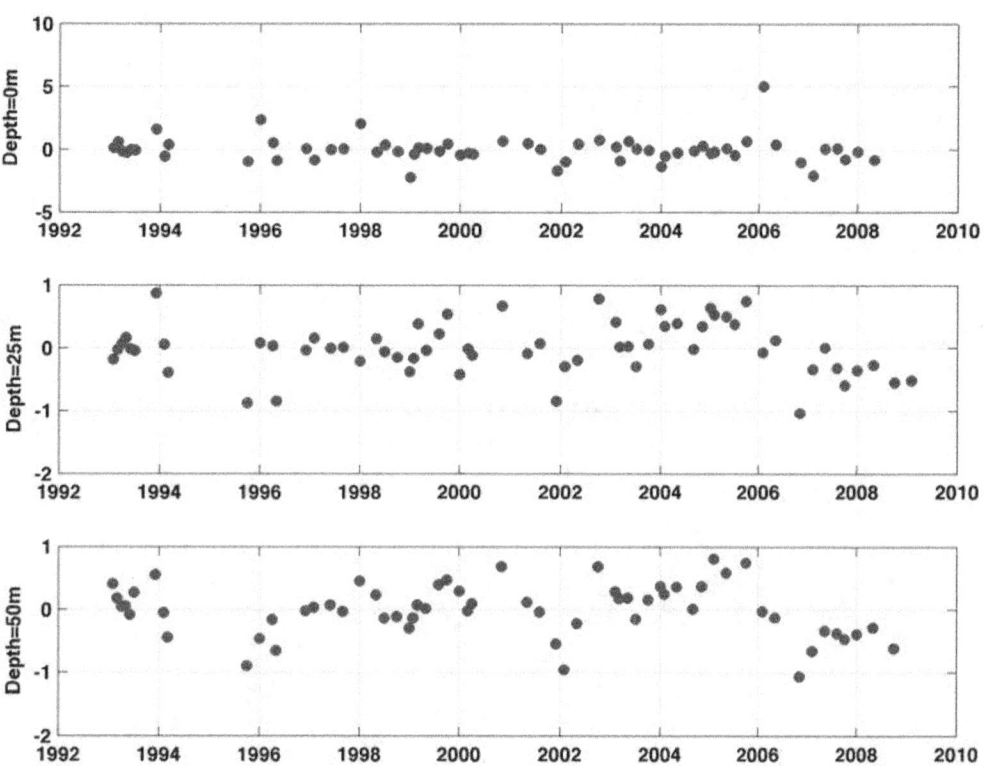

Glacier Bay Oceanographic Monitoring Program Analysis of Observations, 1993–2009
Appendix C. Time series of temperature and salinity anomalies at select depths.

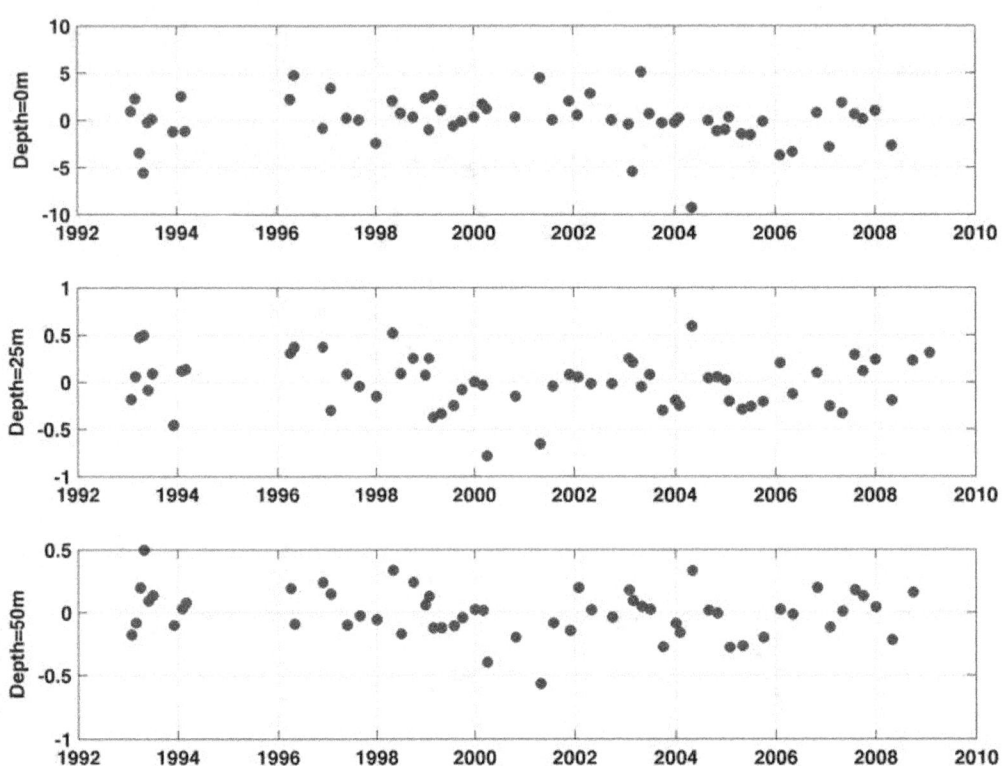

Glacier Bay Oceanographic Monitoring Program Analysis of Observations, 1993–2009
Appendix C. Time series of temperature and salinity anomalies at select depths.

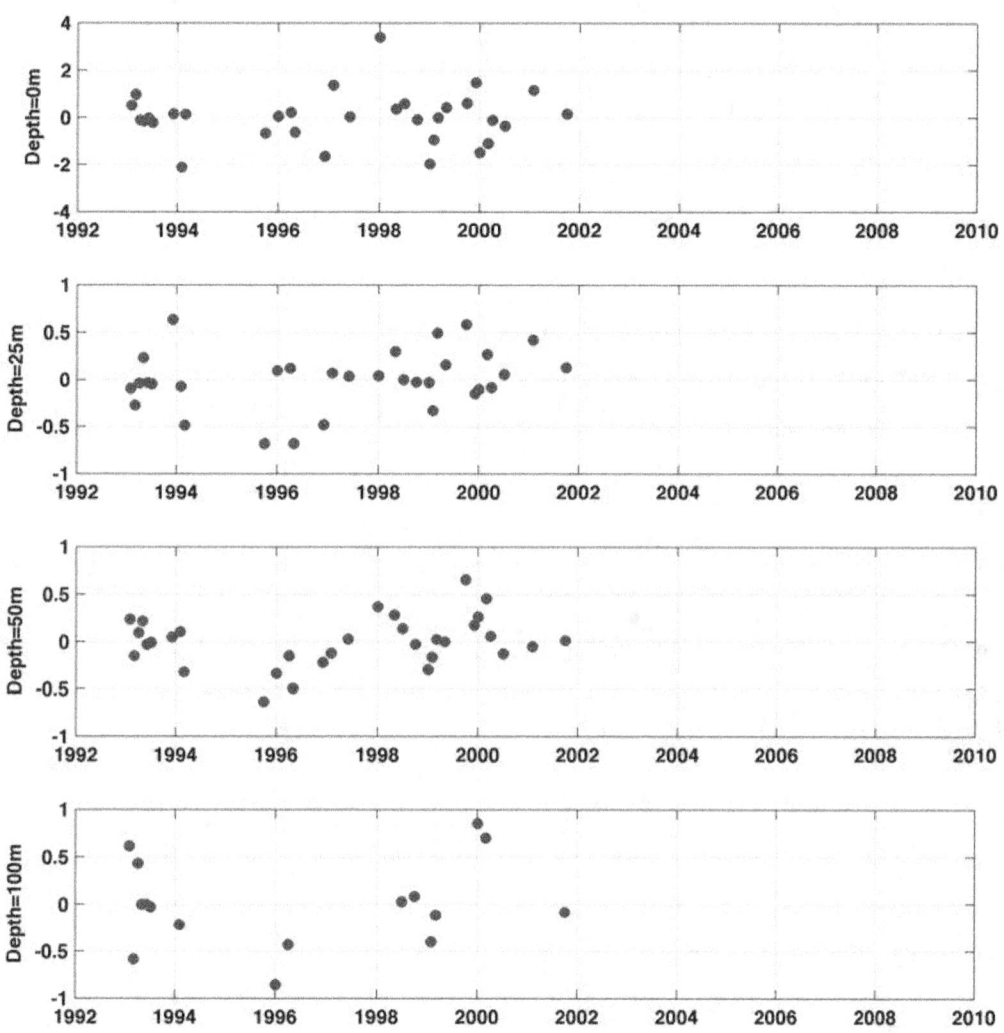

Glacier Bay Oceanographic Monitoring Program Analysis of Observations, 1993–2009
Appendix C. Time series of temperature and salinity anomalies at select depths.

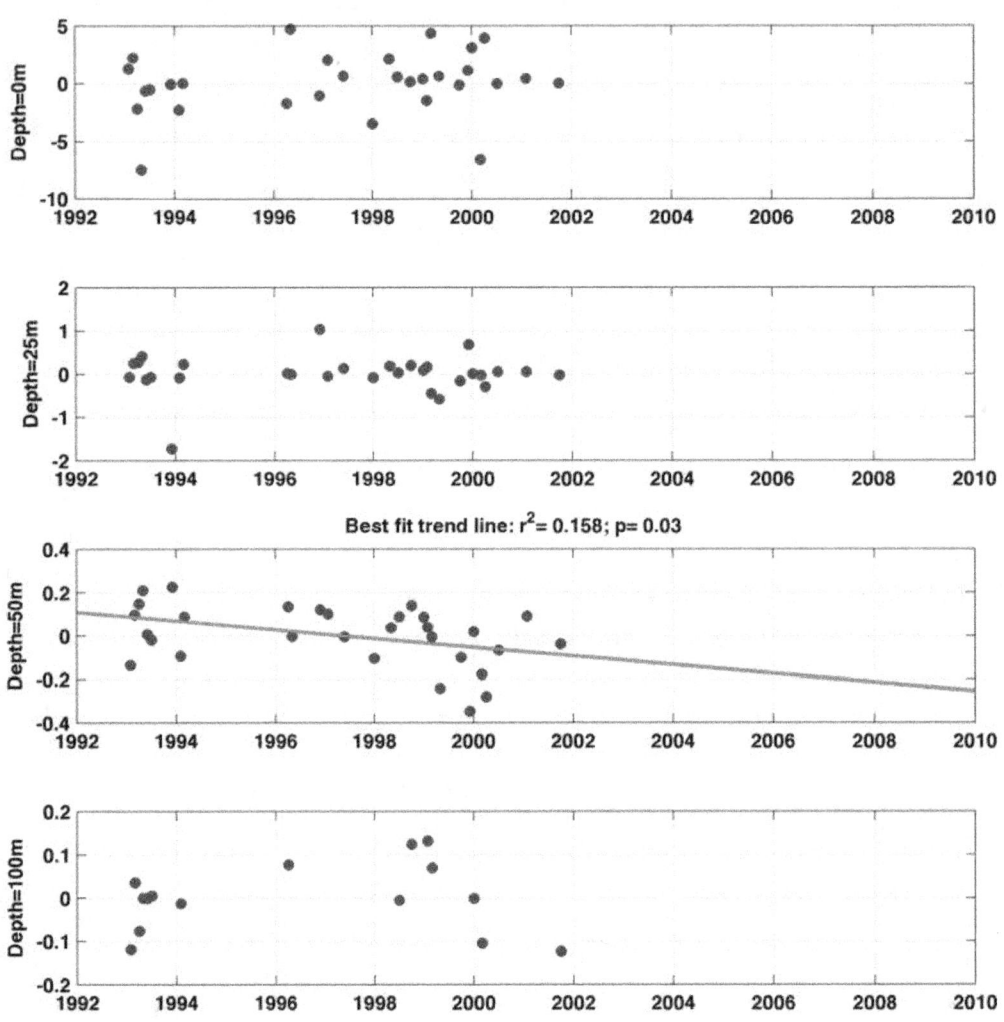

Glacier Bay Oceanographic Monitoring Program Analysis of Observations, 1993–2009
Appendix C. Time series of temperature and salinity anomalies at select depths.

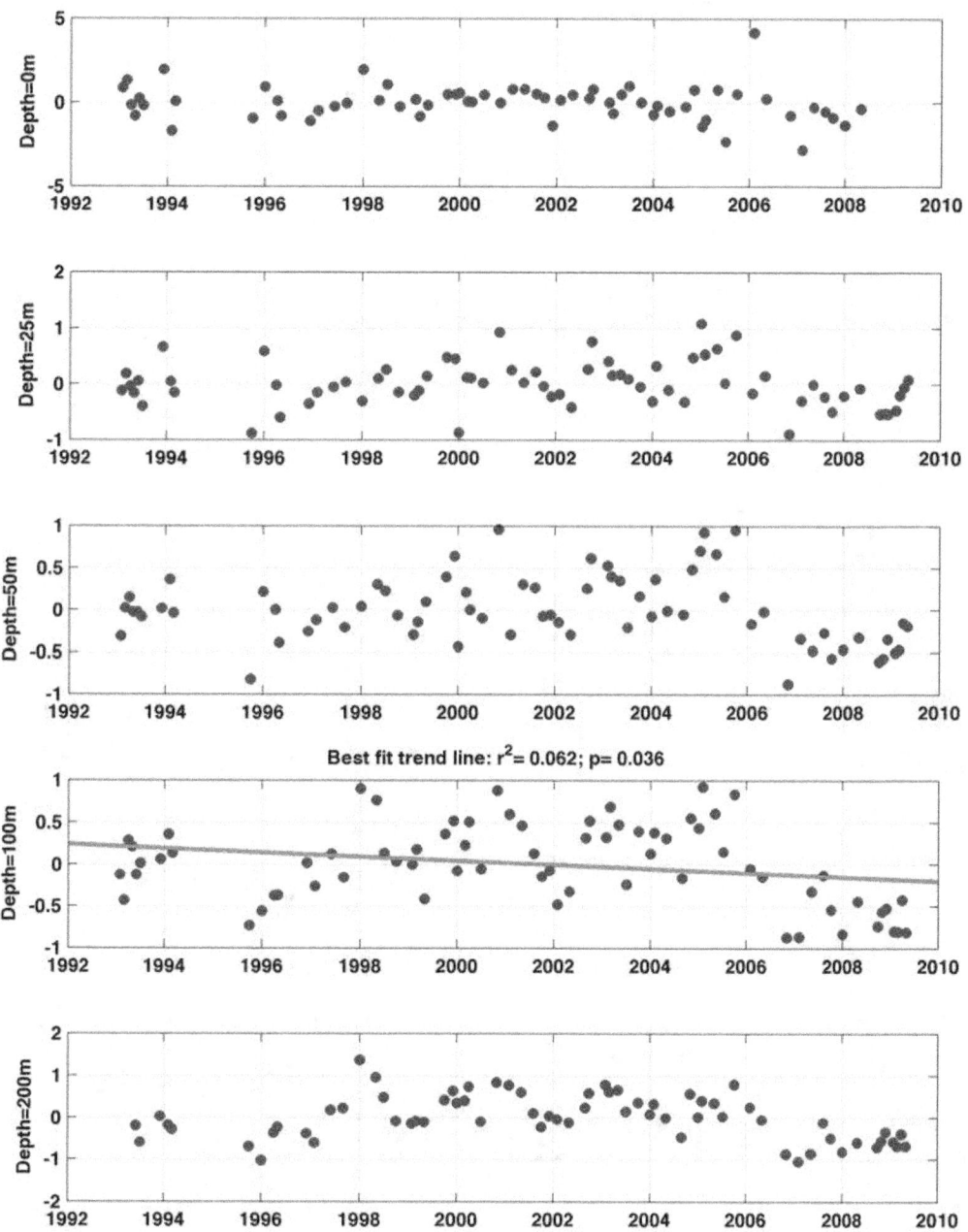

Glacier Bay Oceanographic Monitoring Program Analysis of Observations, 1993–2009
Appendix C. Time series of temperature and salinity anomalies at select depths.

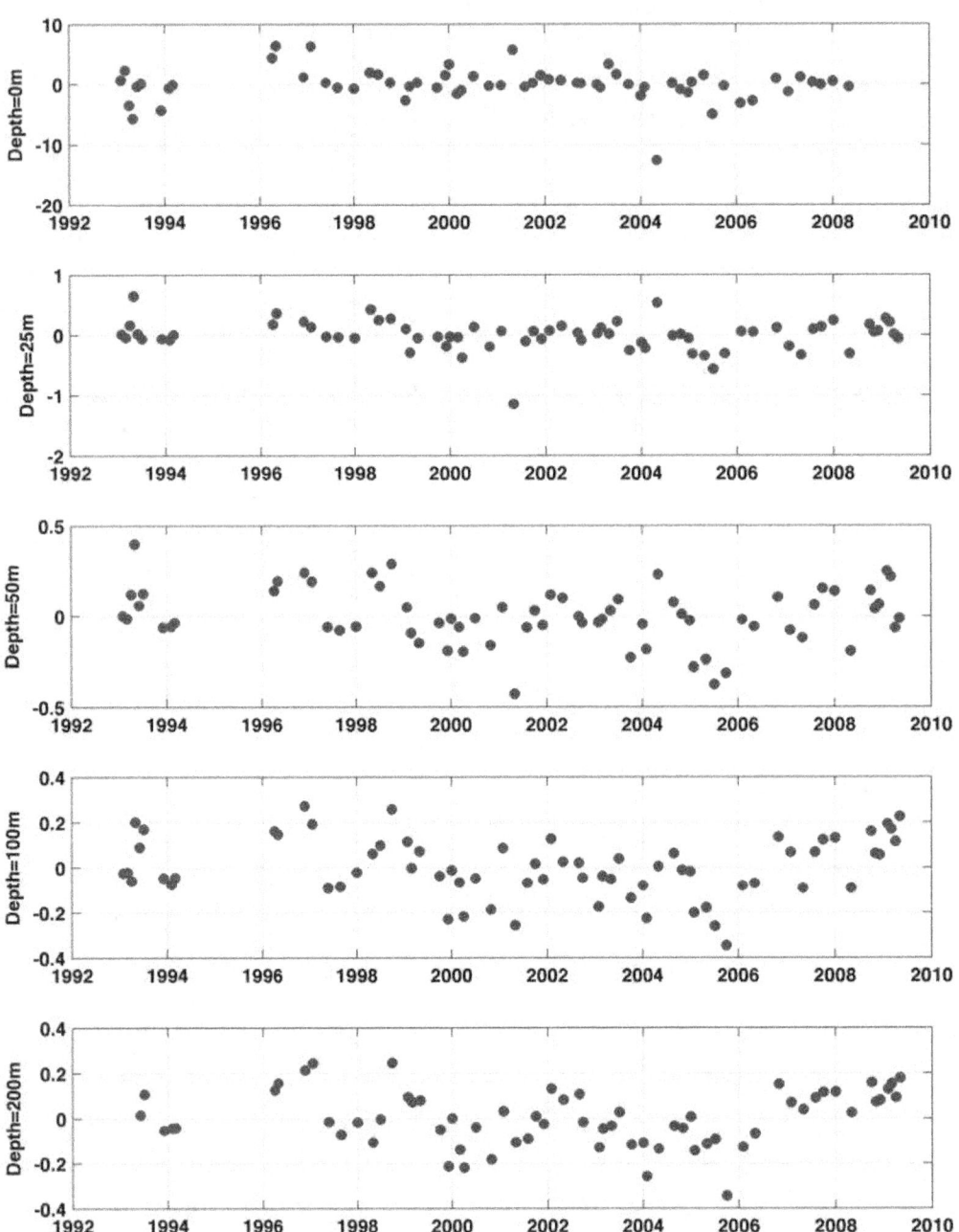

Glacier Bay Oceanographic Monitoring Program Analysis of Observations, 1993–2009
Appendix C. Time series of temperature and salinity anomalies at select depths.

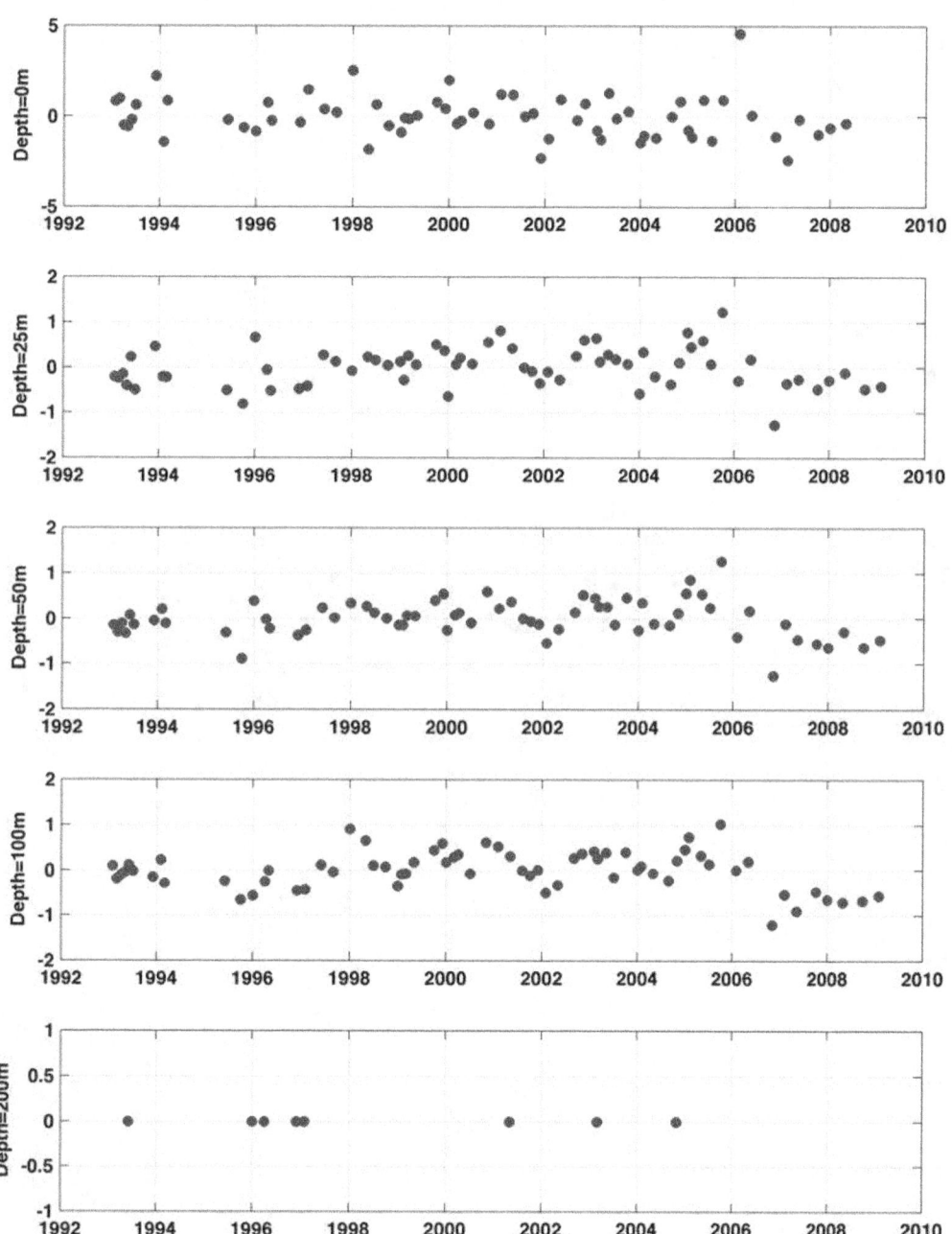

Glacier Bay Oceanographic Monitoring Program Analysis of Observations, 1993–2009
Appendix C. Time series of temperature and salinity anomalies at select depths.

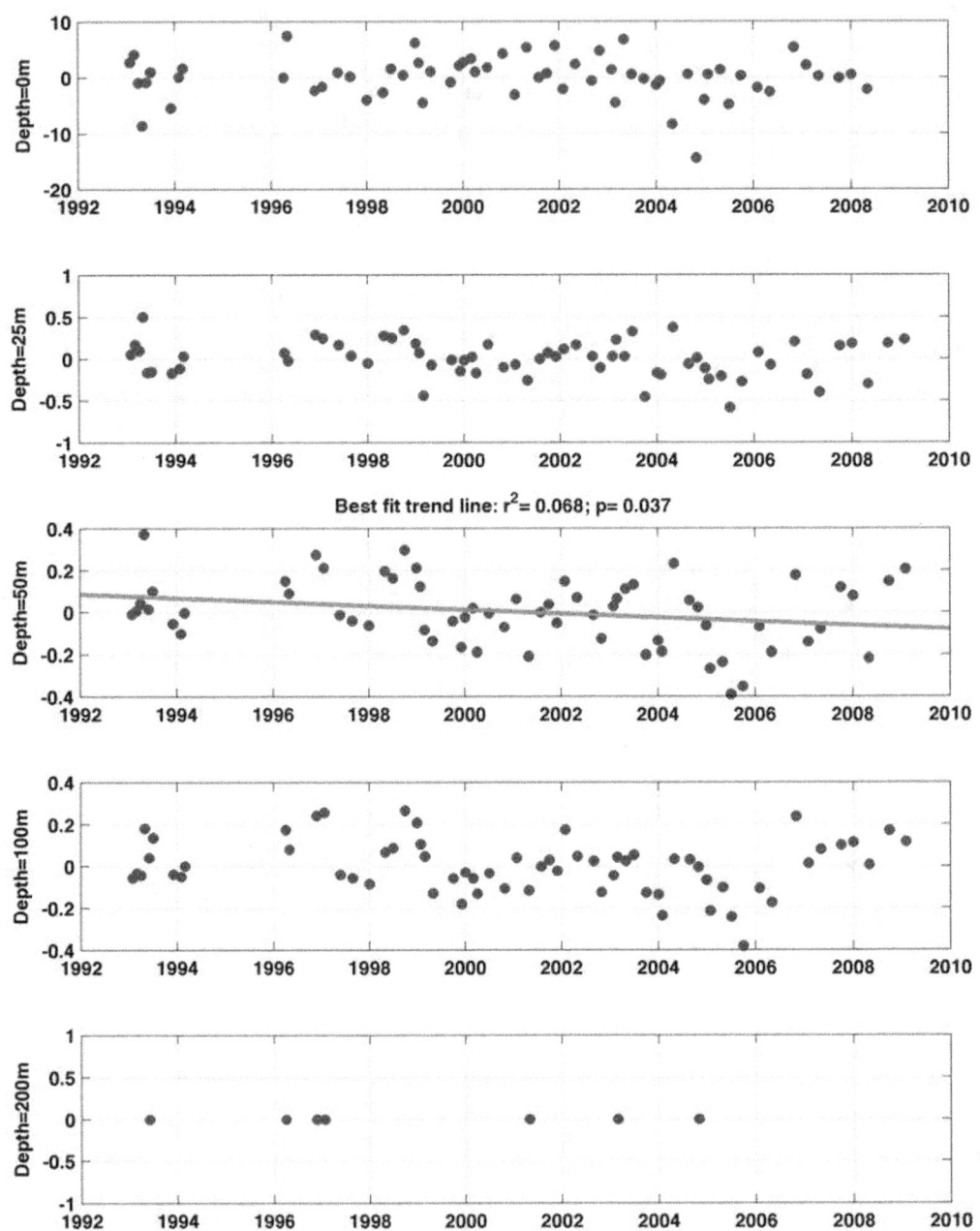

Glacier Bay Oceanographic Monitoring Program Analysis of Observations, 1993–2009
Appendix C. Time series of temperature and salinity anomalies at select depths.

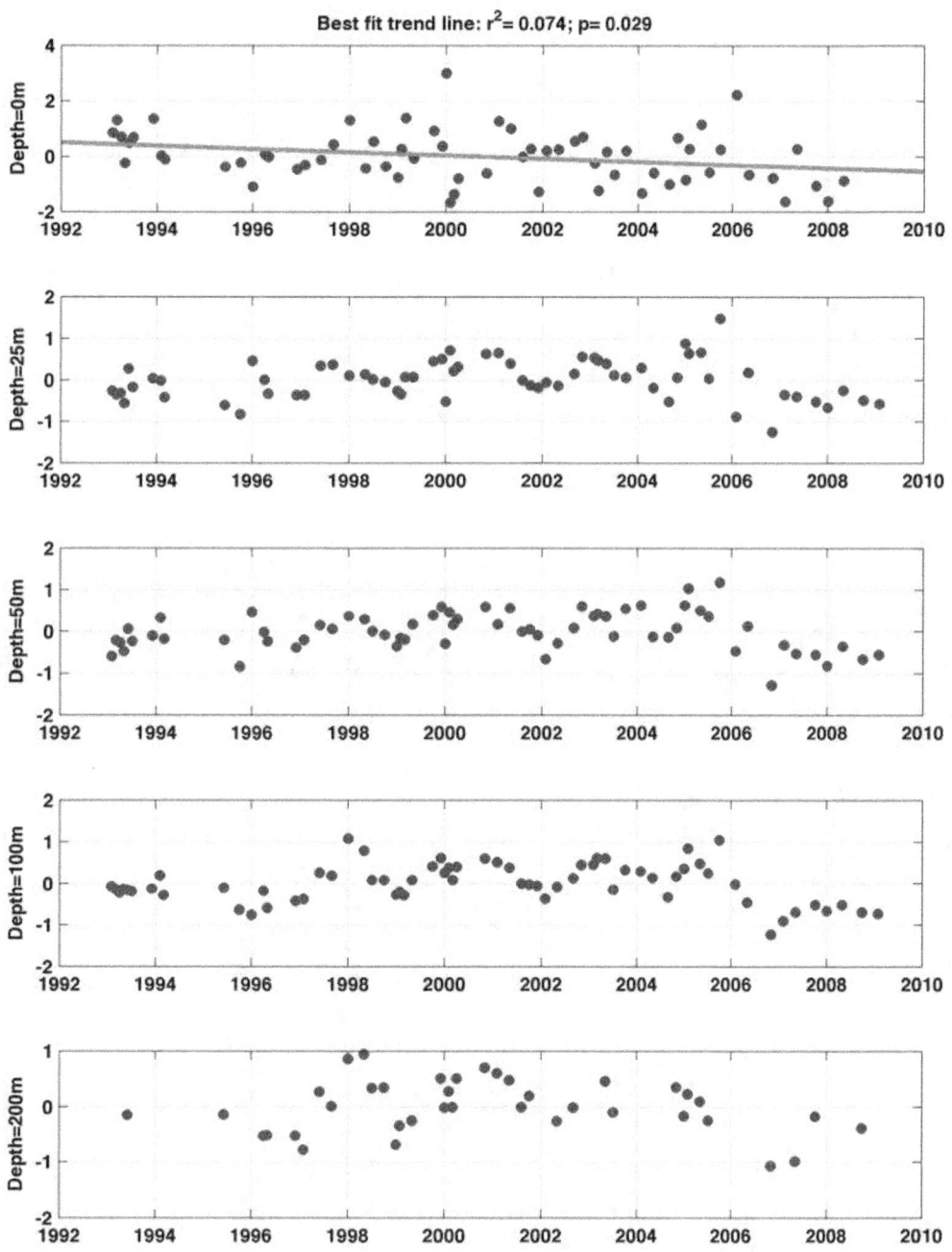

Glacier Bay Oceanographic Monitoring Program Analysis of Observations, 1993–2009
Appendix C. Time series of temperature and salinity anomalies at select depths.

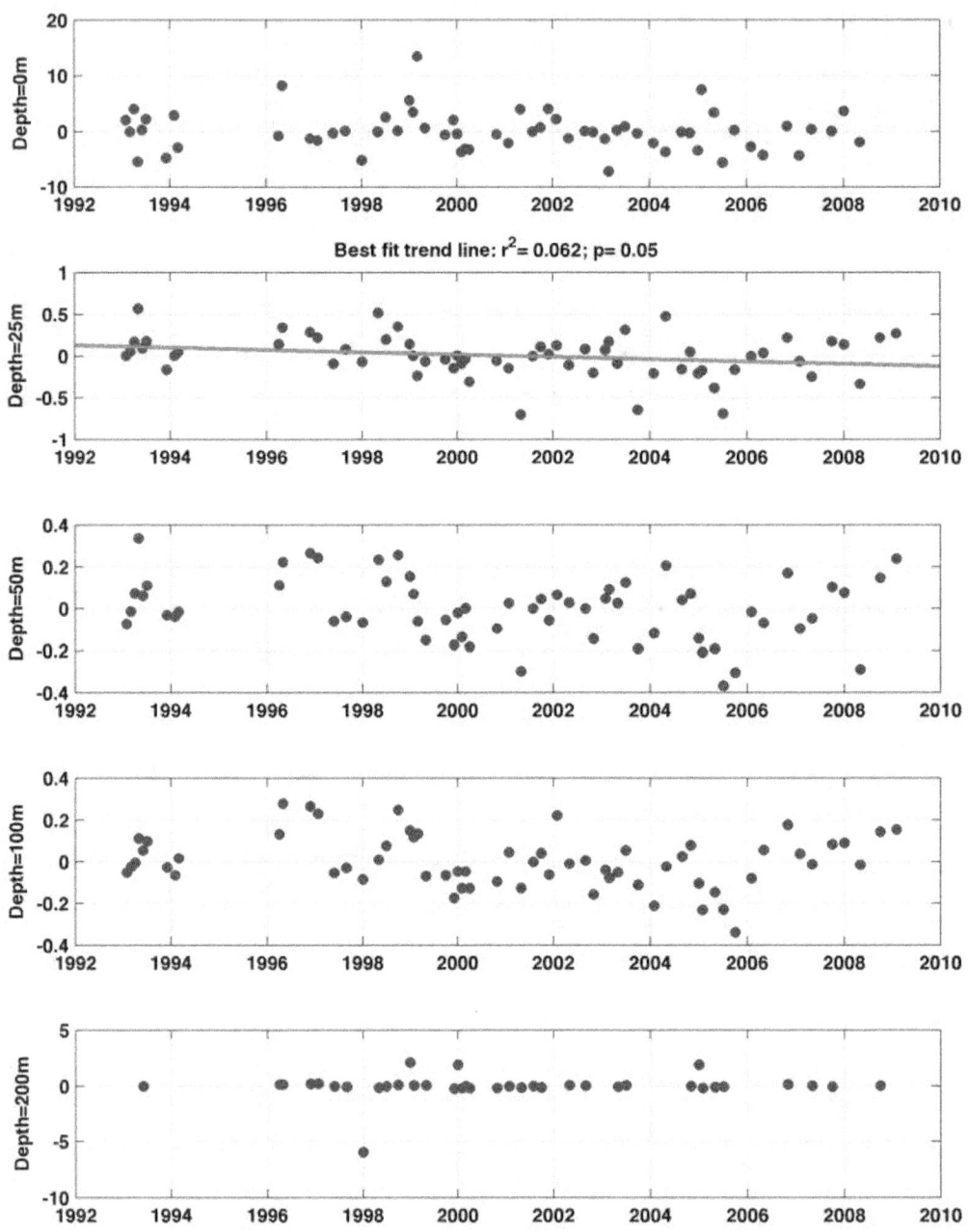

Glacier Bay Oceanographic Monitoring Program Analysis of Observations, 1993–2009
Appendix C. Time series of temperature and salinity anomalies at select depths.

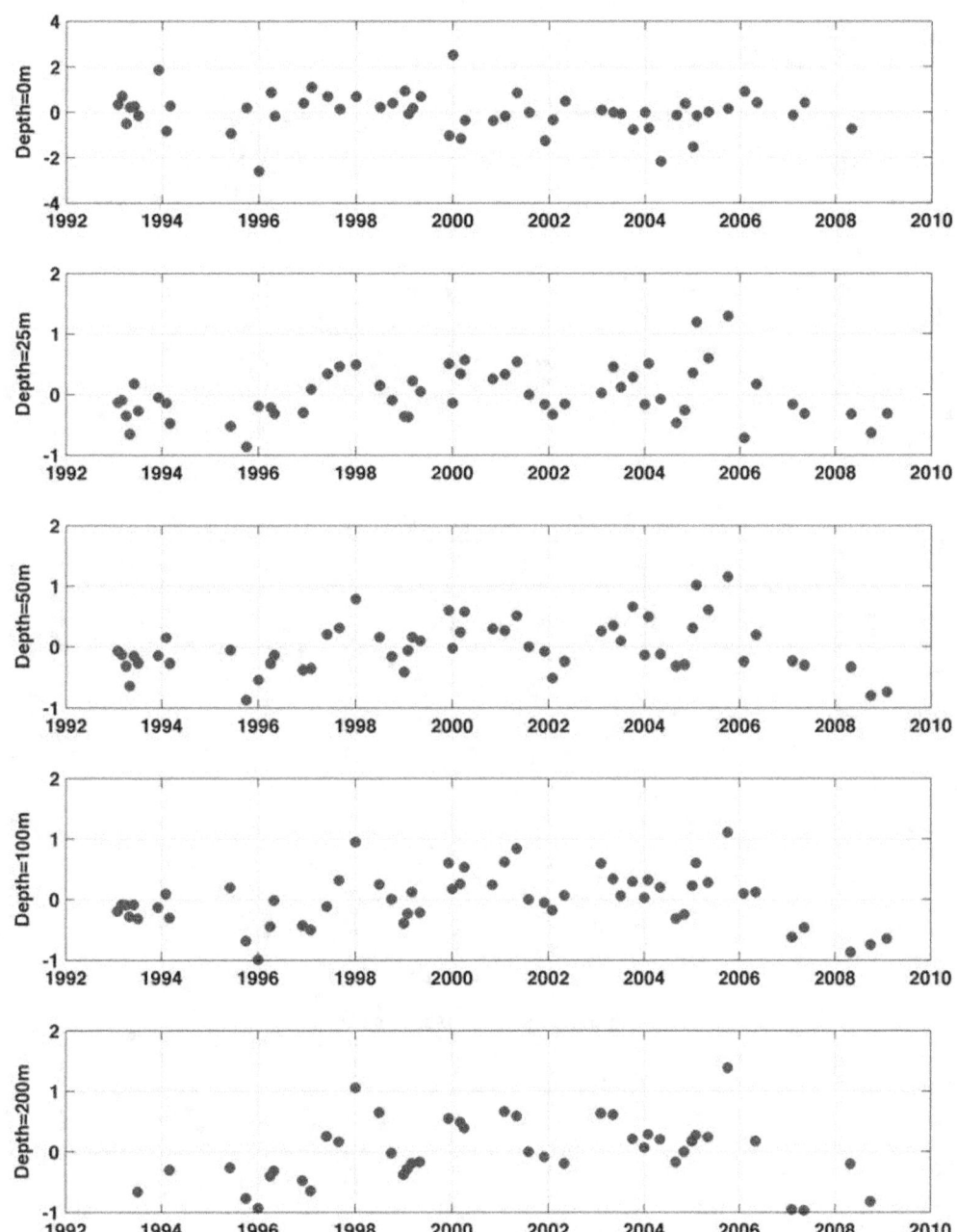

Glacier Bay Oceanographic Monitoring Program Analysis of Observations, 1993–2009
Appendix C. Time series of temperature and salinity anomalies at select depths.

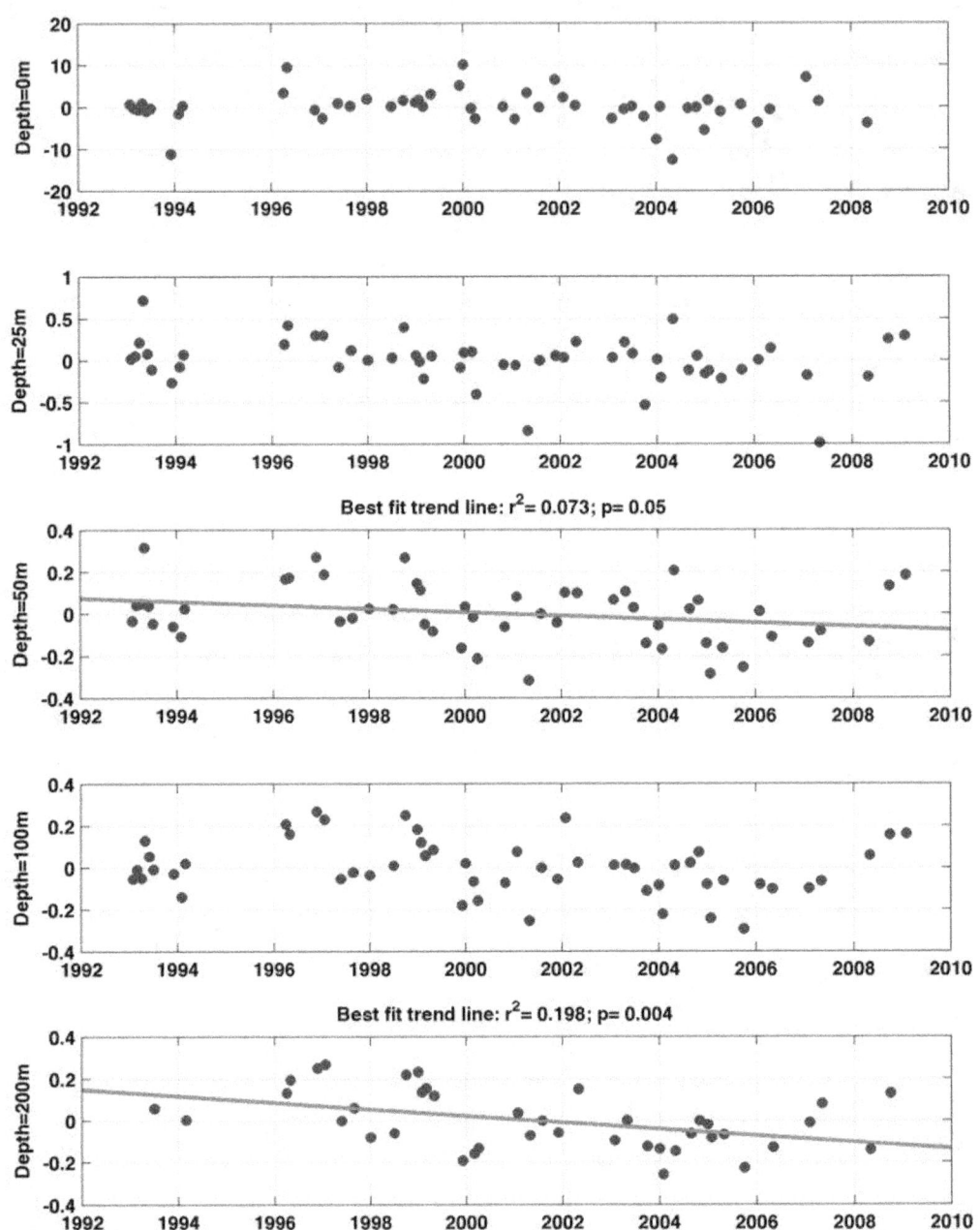

Glacier Bay Oceanographic Monitoring Program Analysis of Observations, 1993–2009
Appendix C. Time series of temperature and salinity anomalies at select depths.

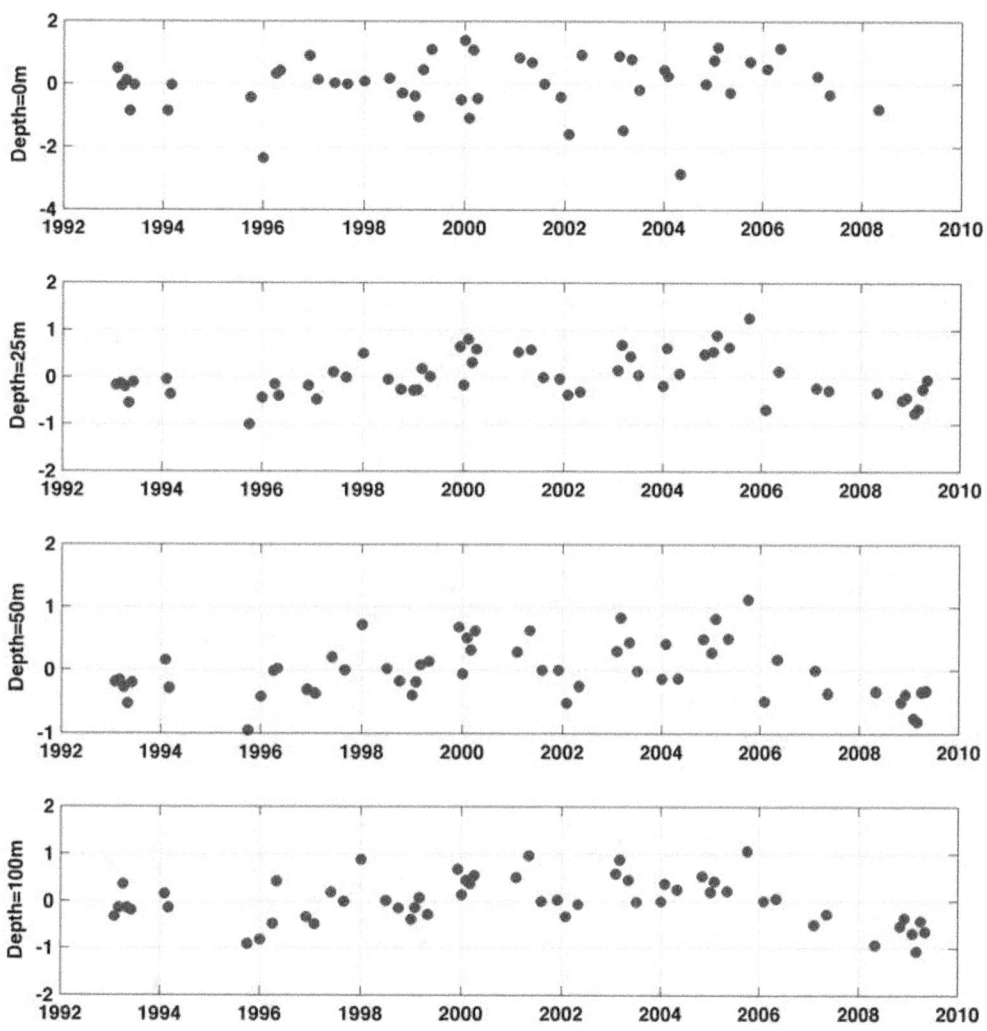

Glacier Bay Oceanographic Monitoring Program Analysis of Observations, 1993–2009
Appendix C. Time series of temperature and salinity anomalies at select depths.

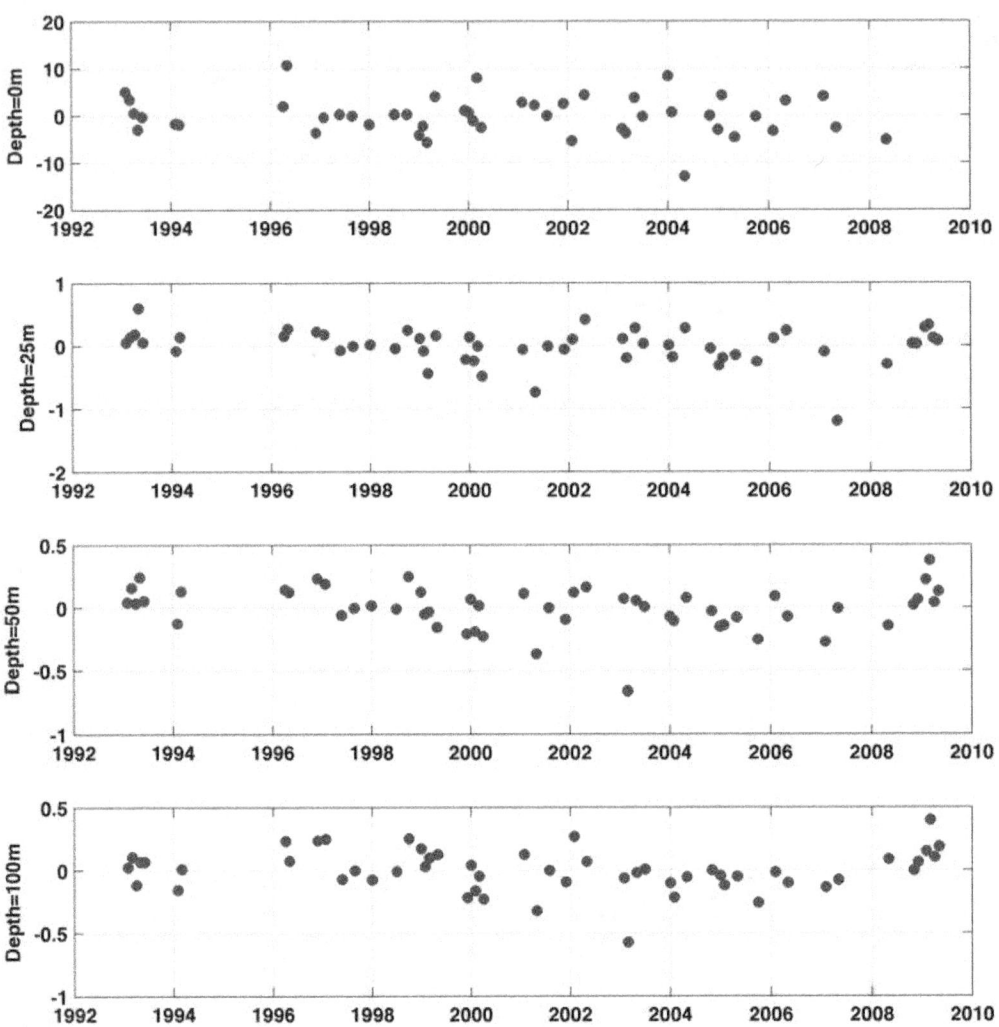

Glacier Bay Oceanographic Monitoring Program Analysis of Observations, 1993–2009
Appendix C. Time series of temperature and salinity anomalies at select depths.

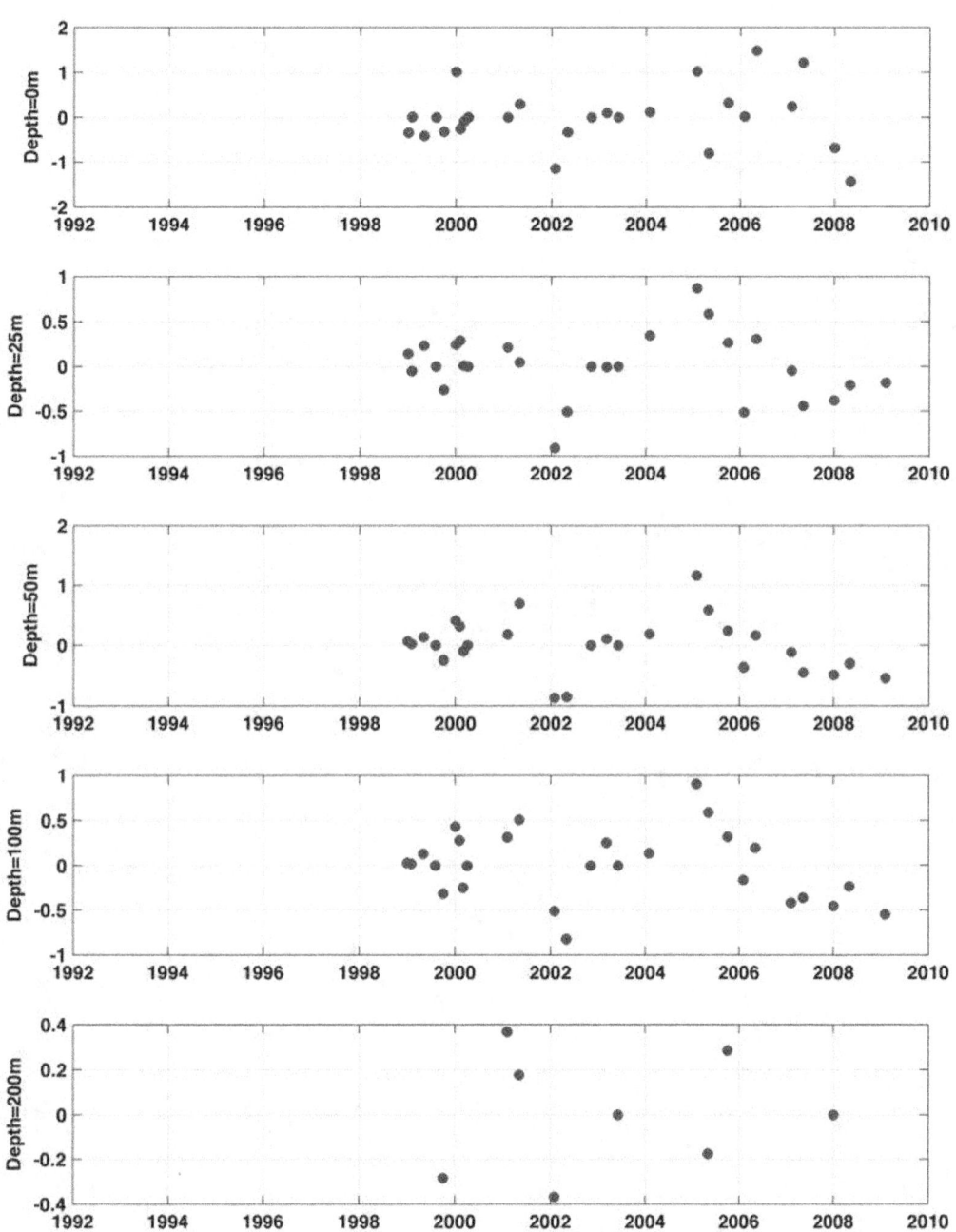

Glacier Bay Oceanographic Monitoring Program Analysis of Observations, 1993–2009
Appendix C. Time series of temperature and salinity anomalies at select depths.

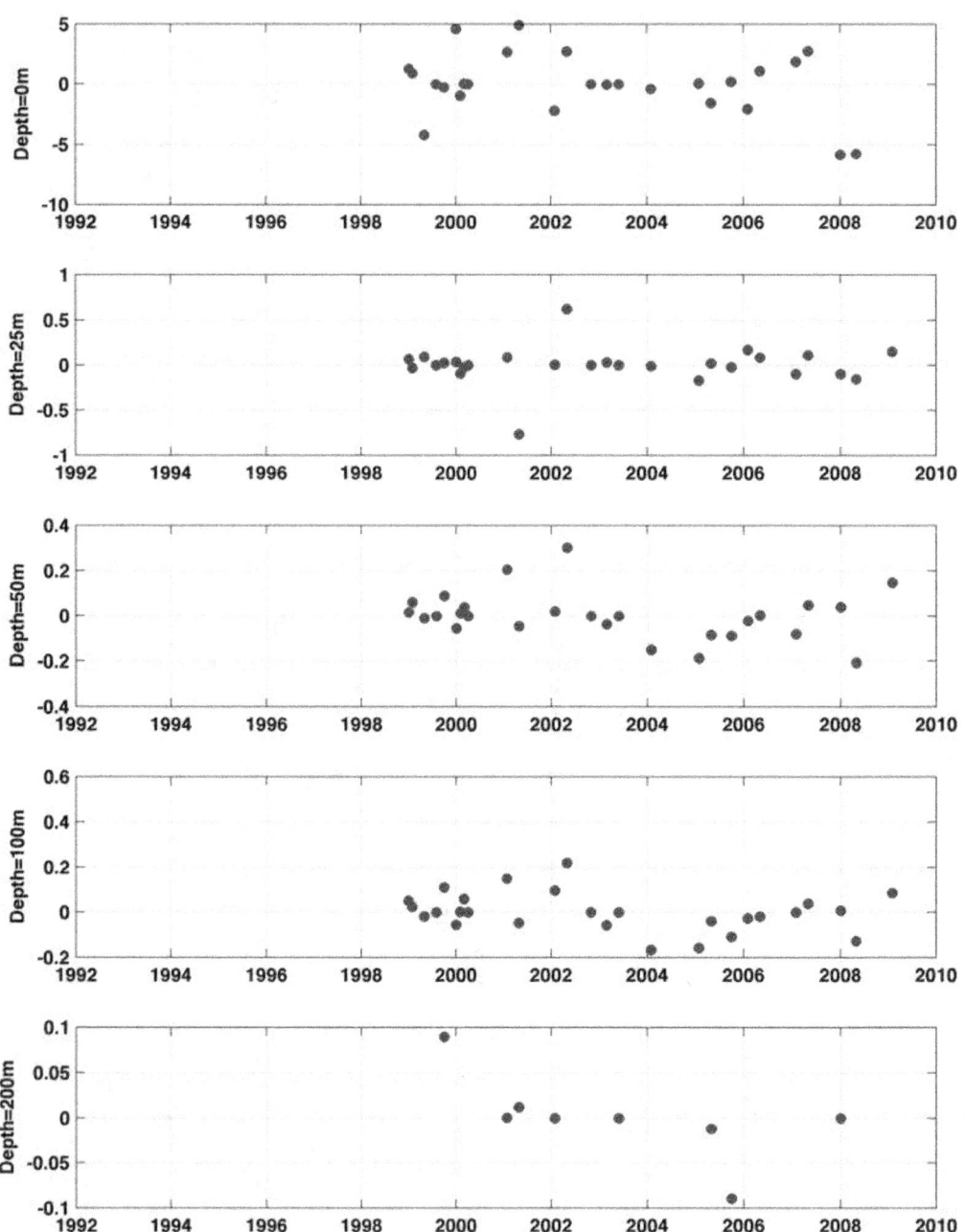

Glacier Bay Oceanographic Monitoring Program Analysis of Observations, 1993–2009
Appendix C. Time series of temperature and salinity anomalies at select depths.

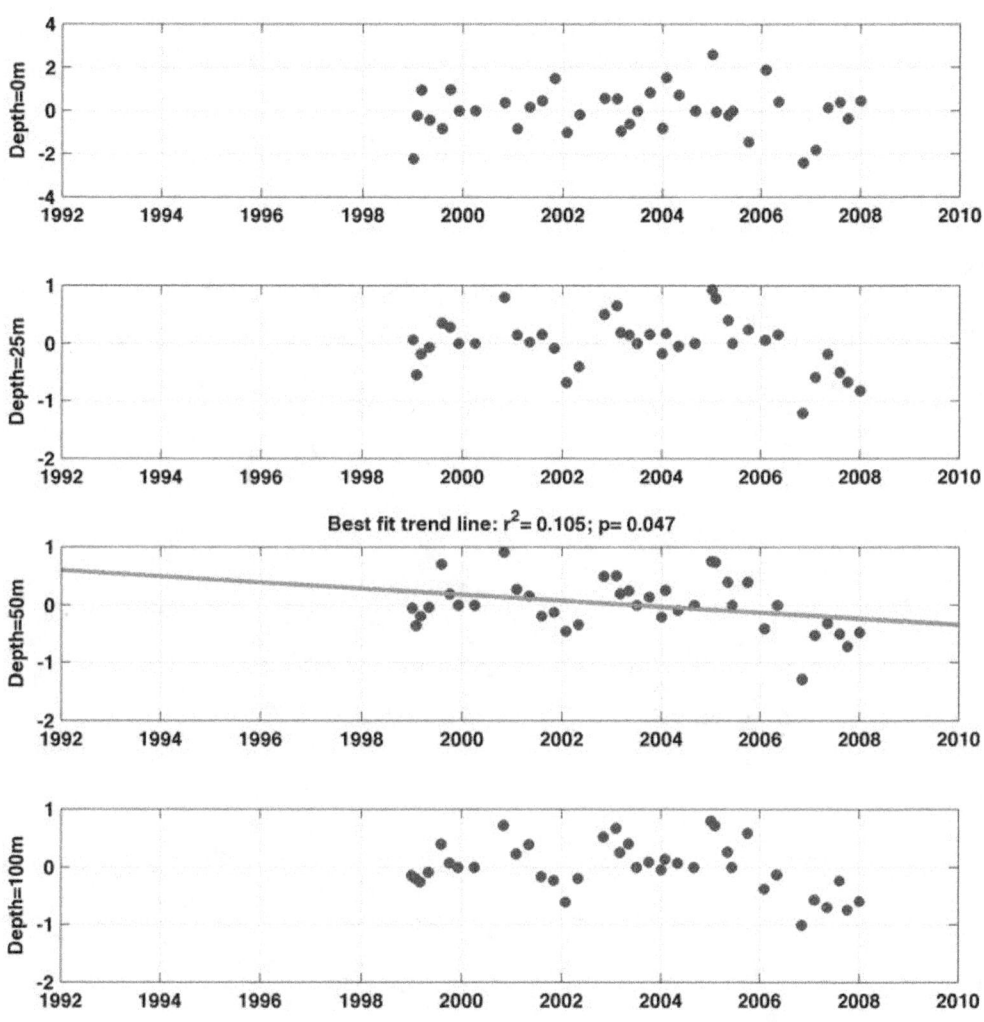

Glacier Bay Oceanographic Monitoring Program Analysis of Observations, 1993–2009
Appendix C. Time series of temperature and salinity anomalies at select depths.

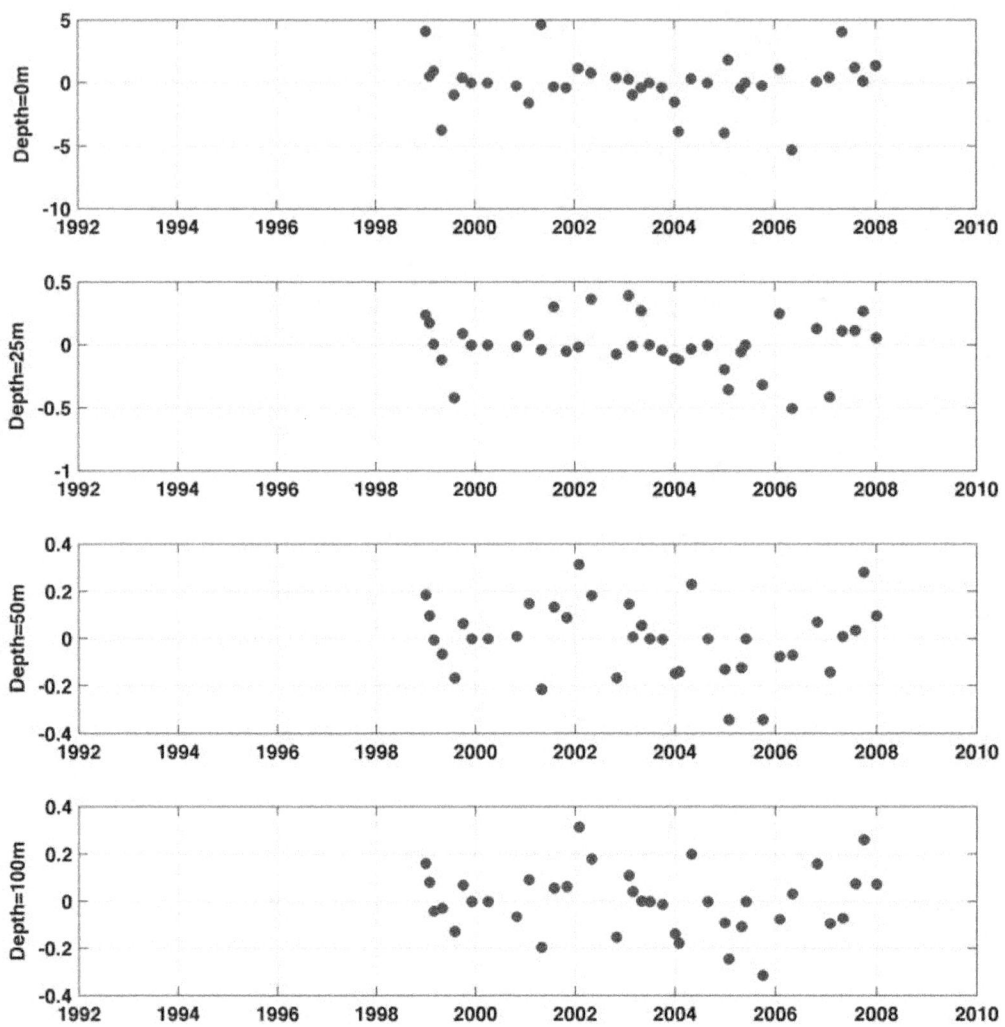

Glacier Bay Oceanographic Monitoring Program Analysis of Observations, 1993–2009
Appendix C. Time series of temperature and salinity anomalies at select depths.

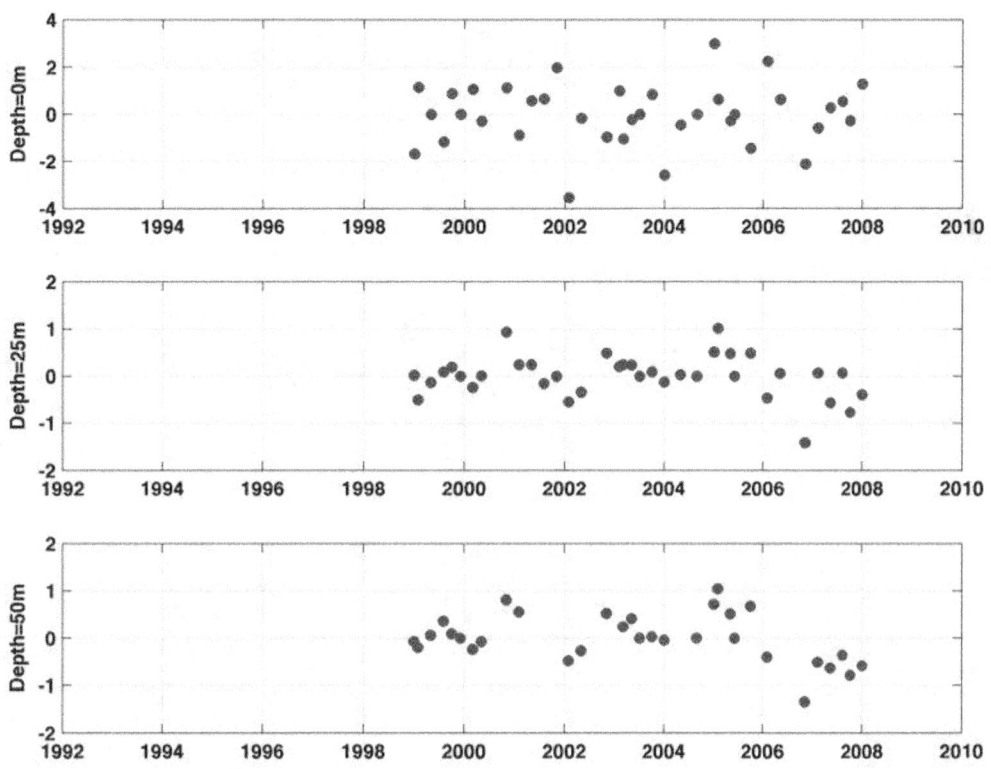

Glacier Bay Oceanographic Monitoring Program Analysis of Observations, 1993–2009
Appendix C. Time series of temperature and salinity anomalies at select depths.

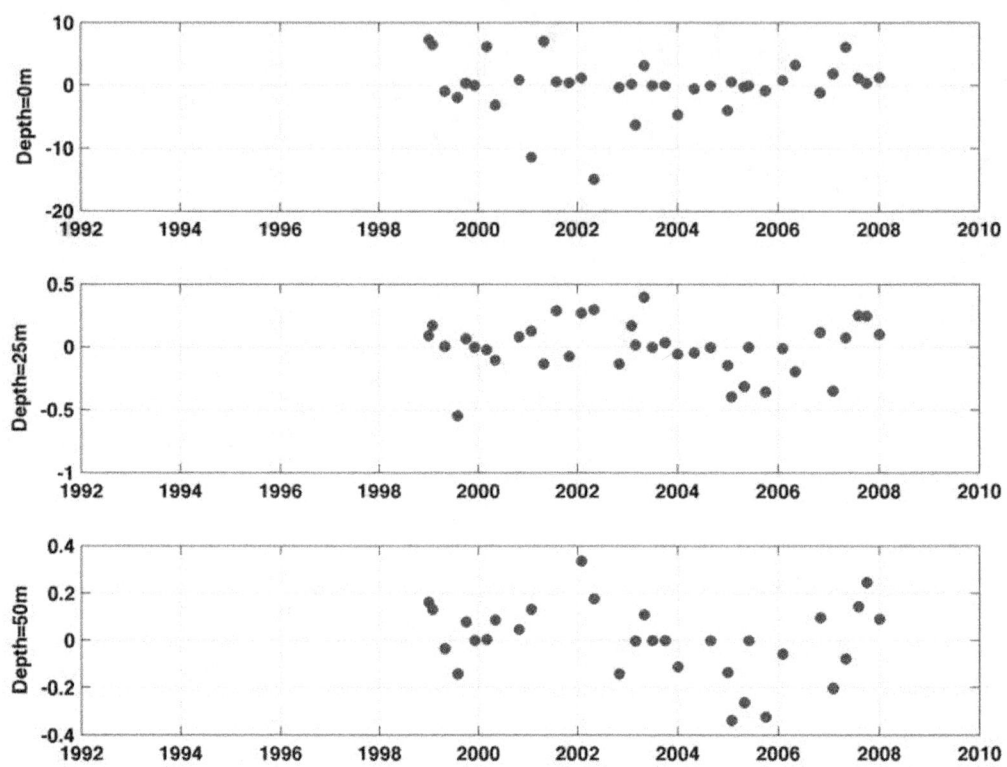

Appendix D. Temperature and salinity anomalies at select depths.

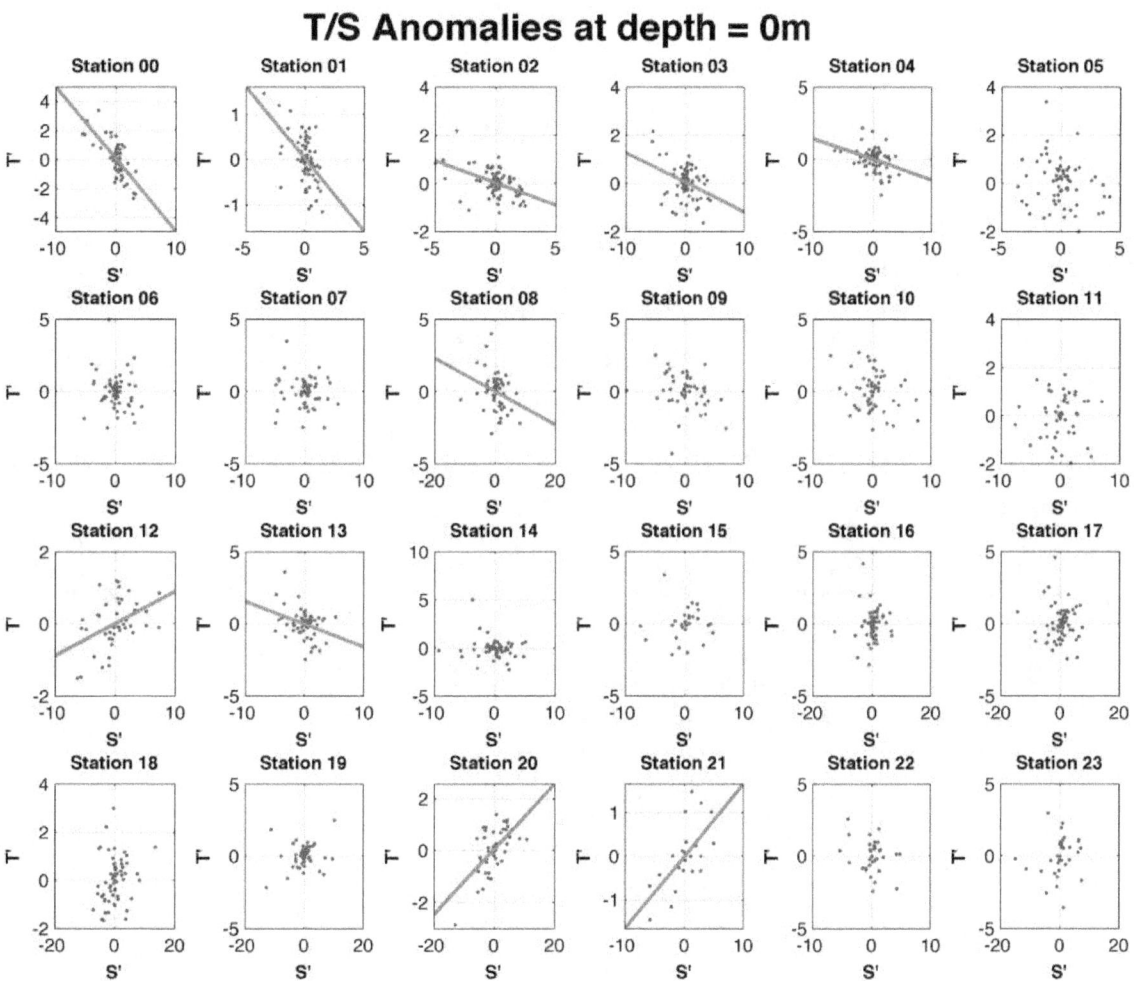

Glacier Bay Oceanographic Monitoring Program Analysis of Observations, 1993–2009
Appendix D. Temperature and salinity anomalies at select depths.

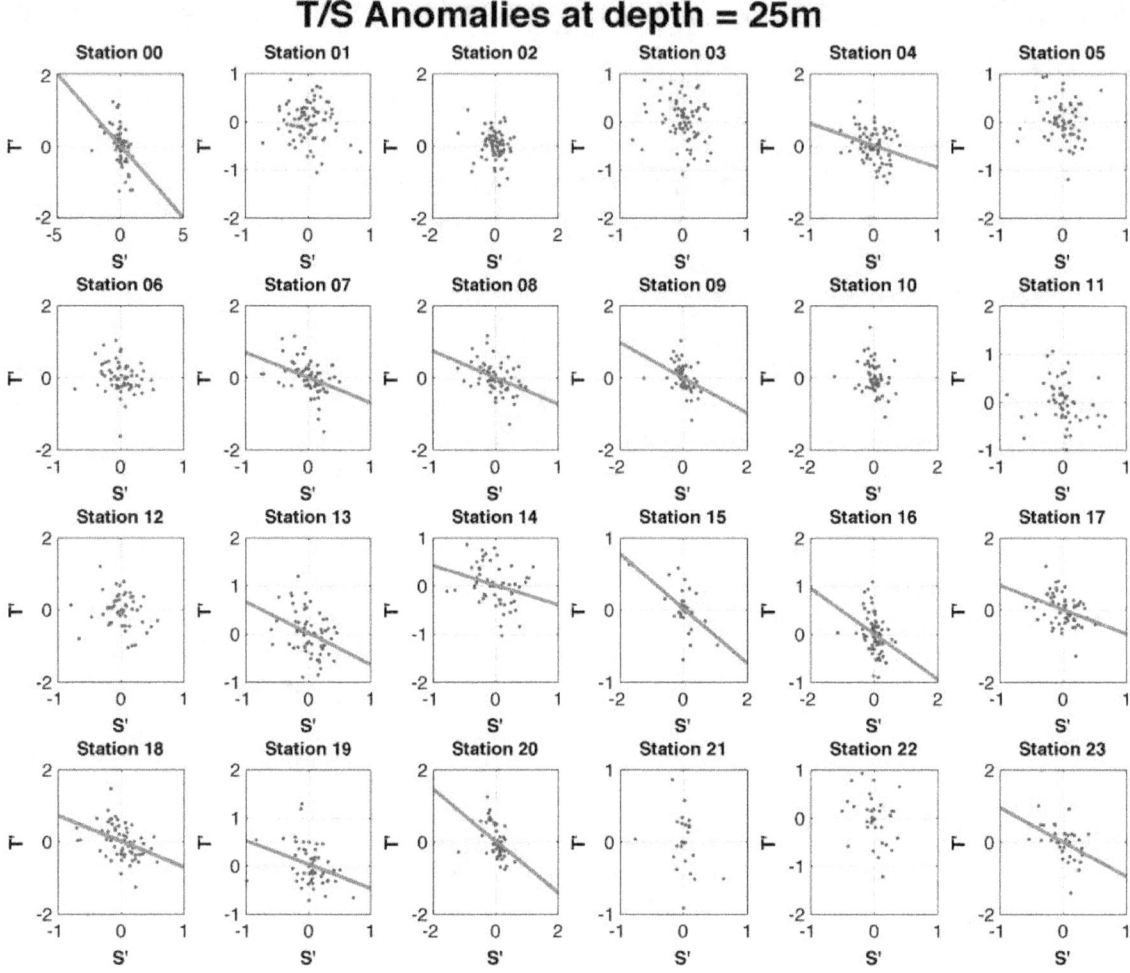

Glacier Bay Oceanographic Monitoring Program Analysis of Observations, 1993–2009
Appendix D. Temperature and salinity anomalies at select depths.

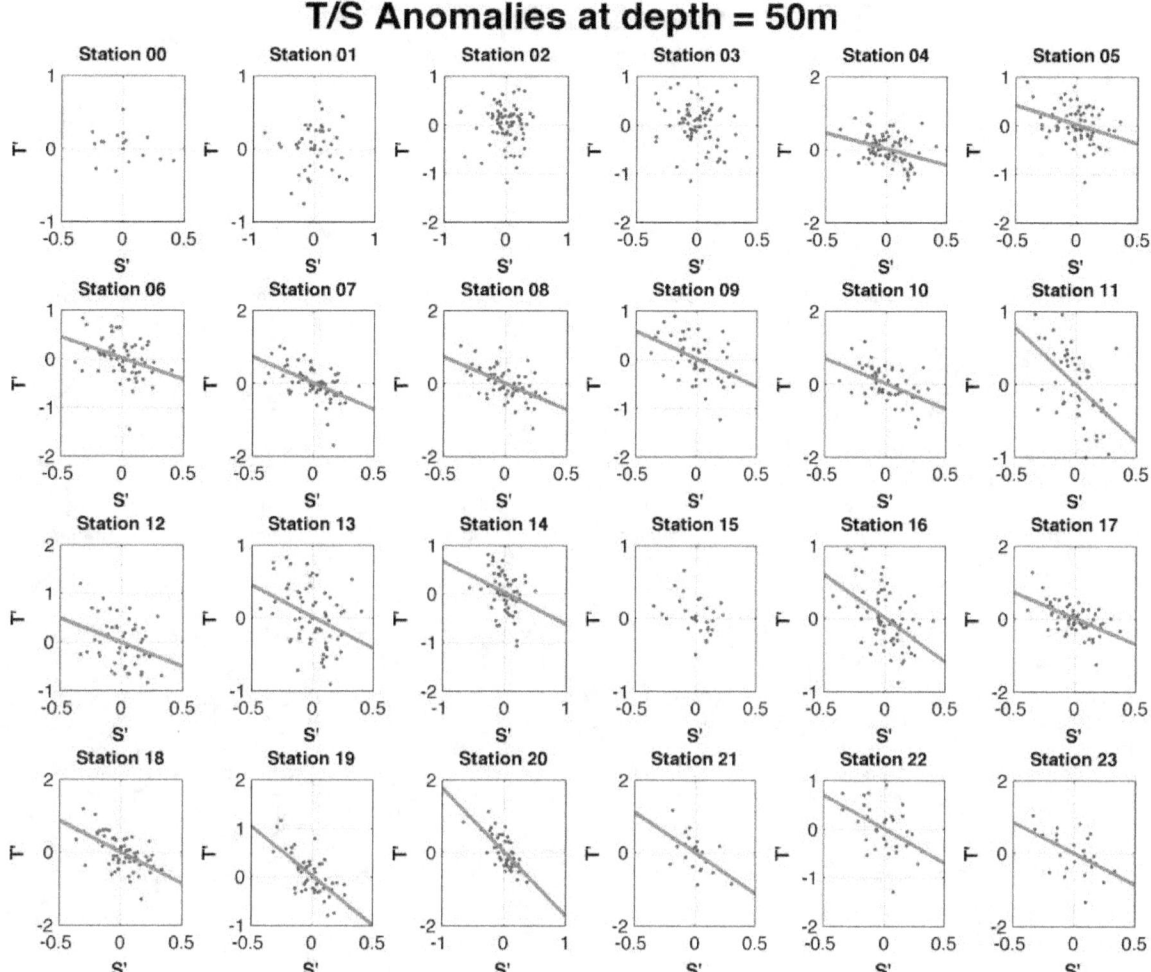

Glacier Bay Oceanographic Monitoring Program Analysis of Observations, 1993–2009
Appendix D. Temperature and salinity anomalies at select depths.

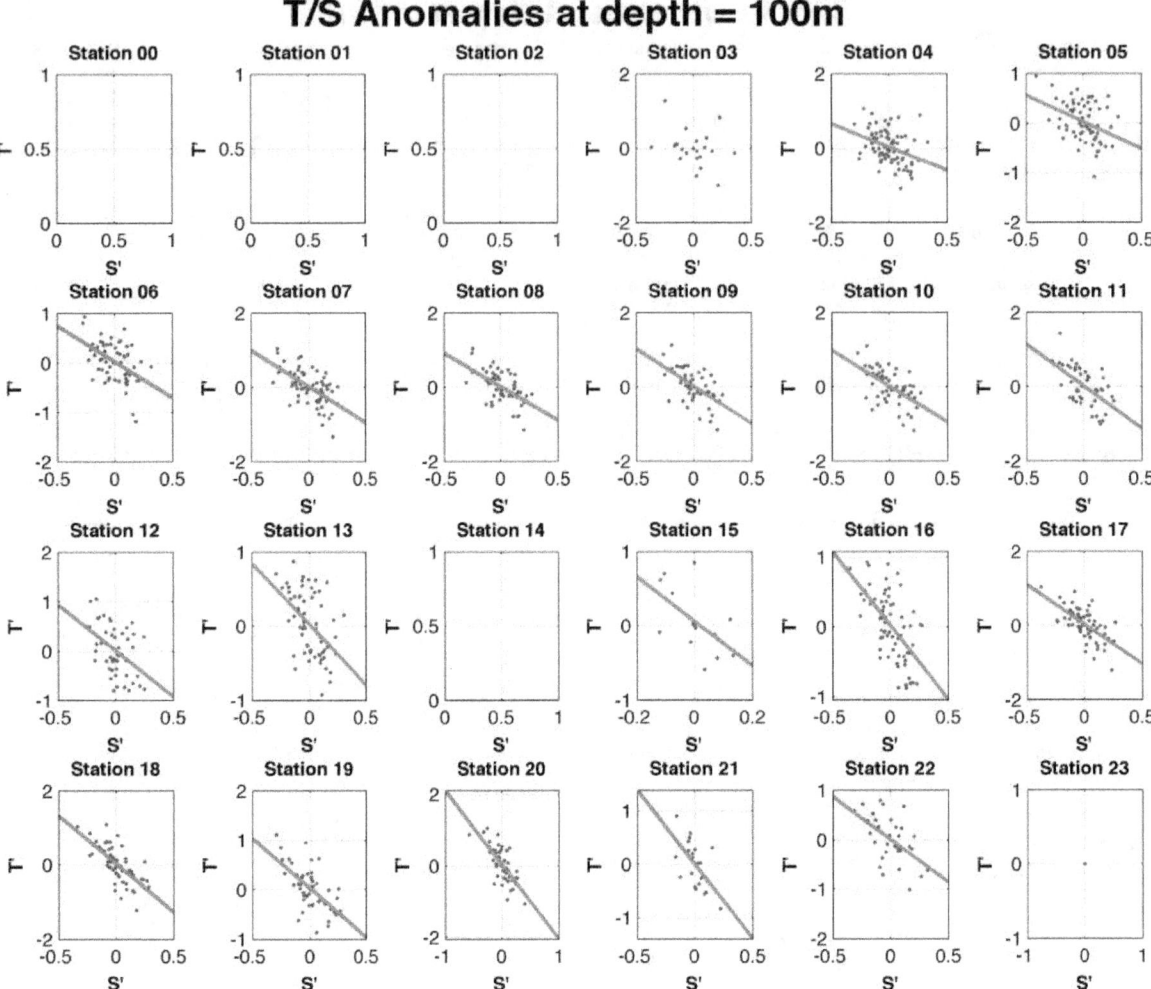

Glacier Bay Oceanographic Monitoring Program Analysis of Observations, 1993–2009
Appendix D. Temperature and salinity anomalies at select depths.

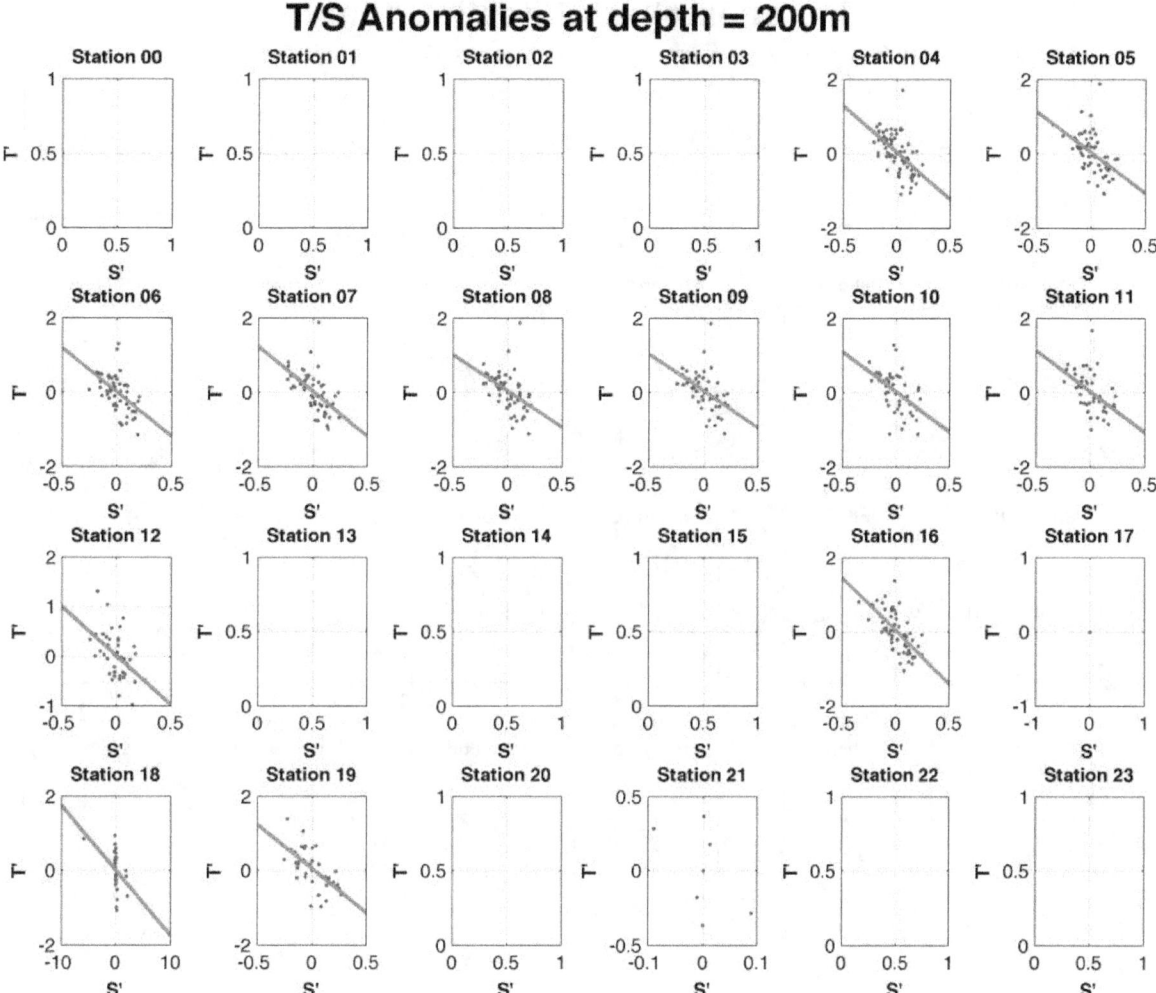

Glacier Bay Oceanographic Monitoring Program Analysis of Observations, 1993–2009
Appendix D. Temperature and salinity anomalies at select depths.

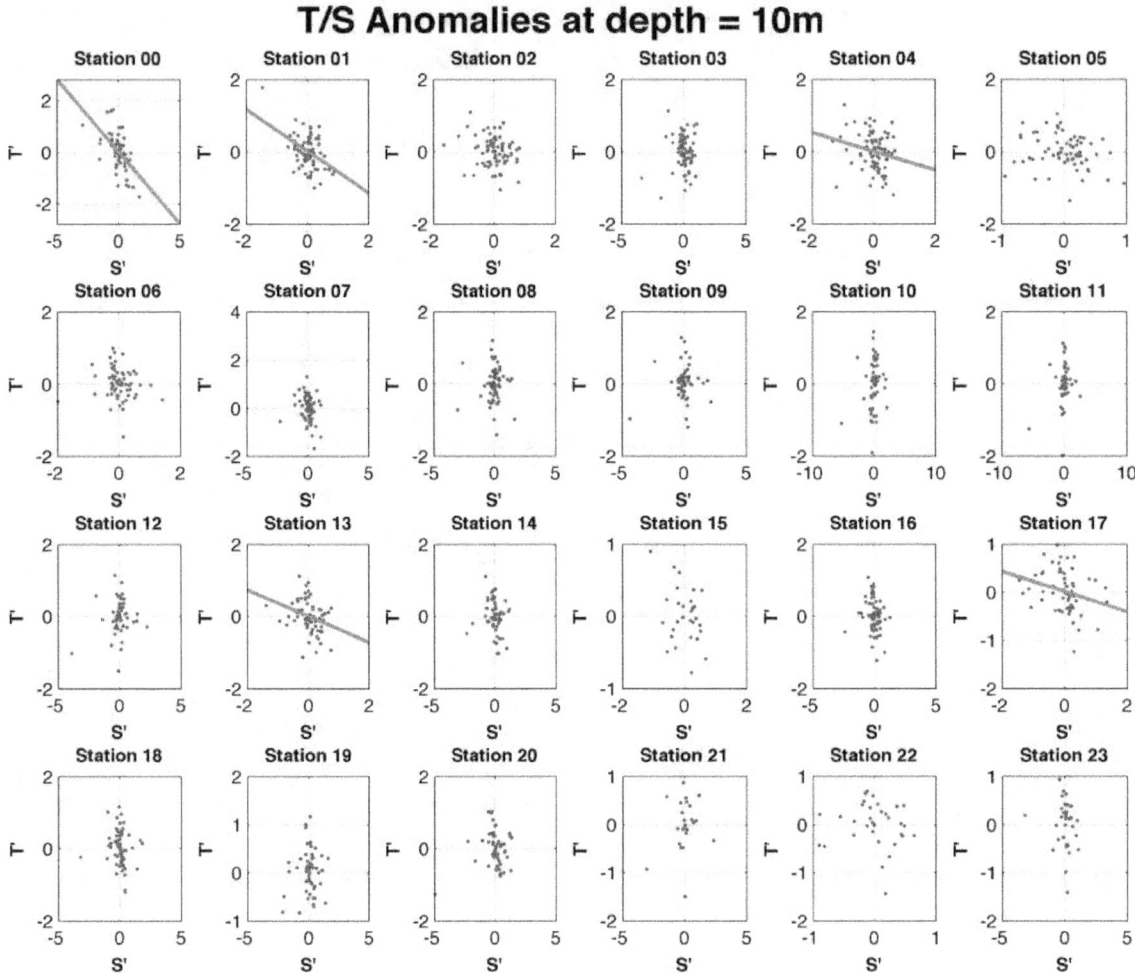

Appendix E. Regressions of temperature and salinity anomalies with atmospheric, oceanographic and climate time series.

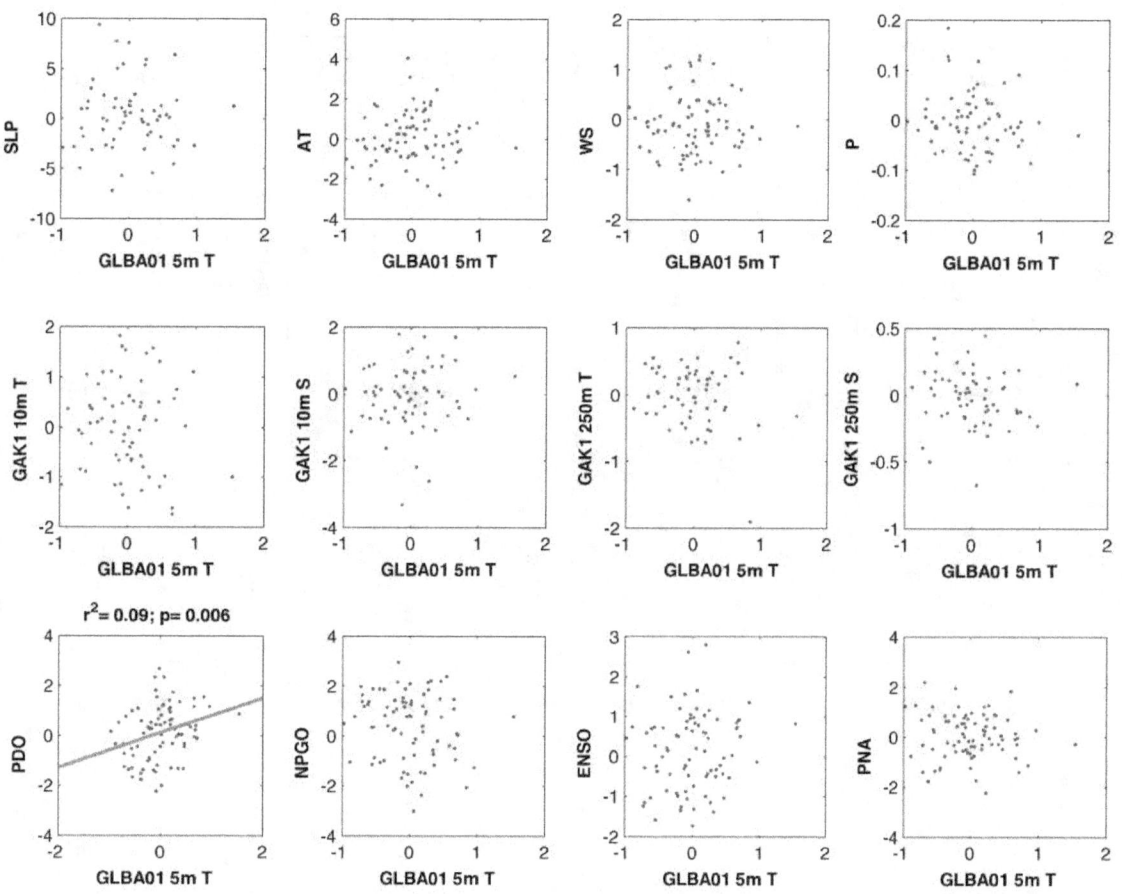

Glacier Bay Oceanographic Monitoring Program Analysis of Observations, 1993–2009
Appendix E. Regressions of temperature and salinity anomalies with atmospheric, oceanographic and climate time series.

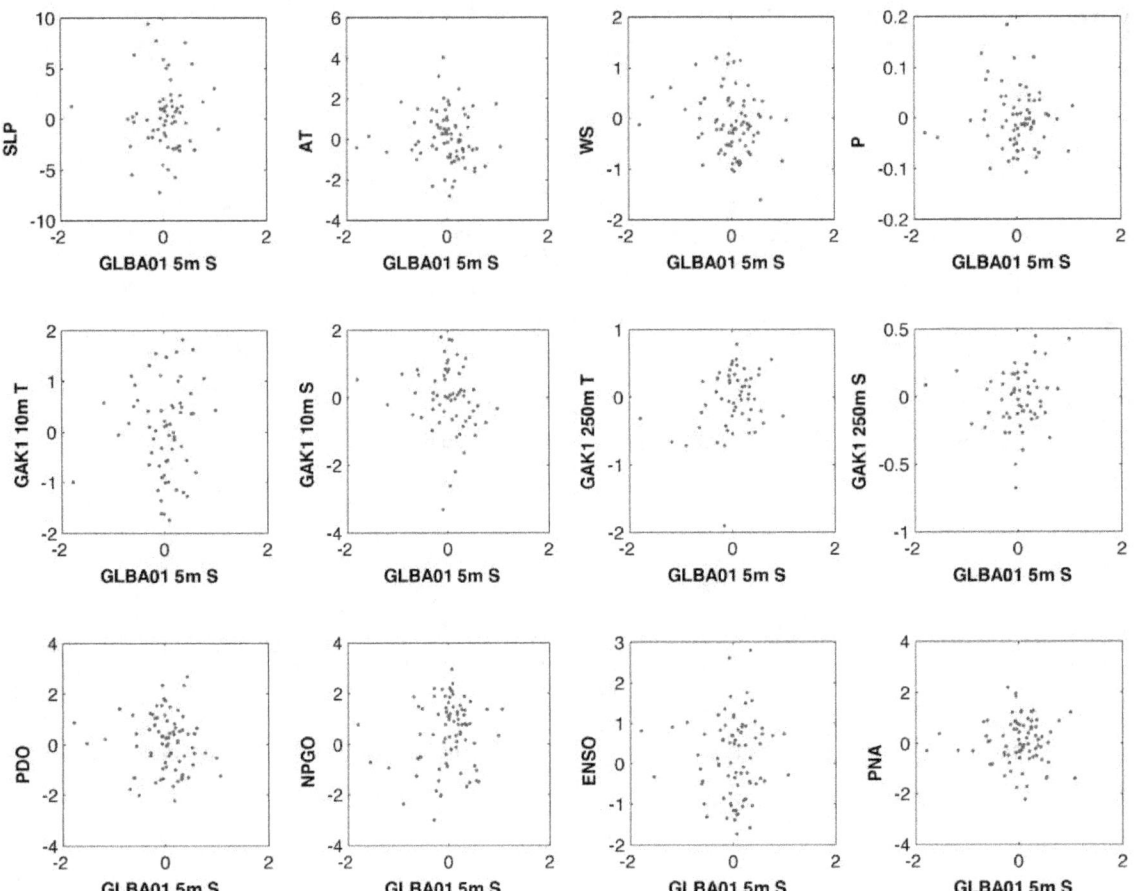

Glacier Bay Oceanographic Monitoring Program Analysis of Observations, 1993–2009
Appendix E. Regressions of temperature and salinity anomalies with atmospheric, oceanographic and climate time series.

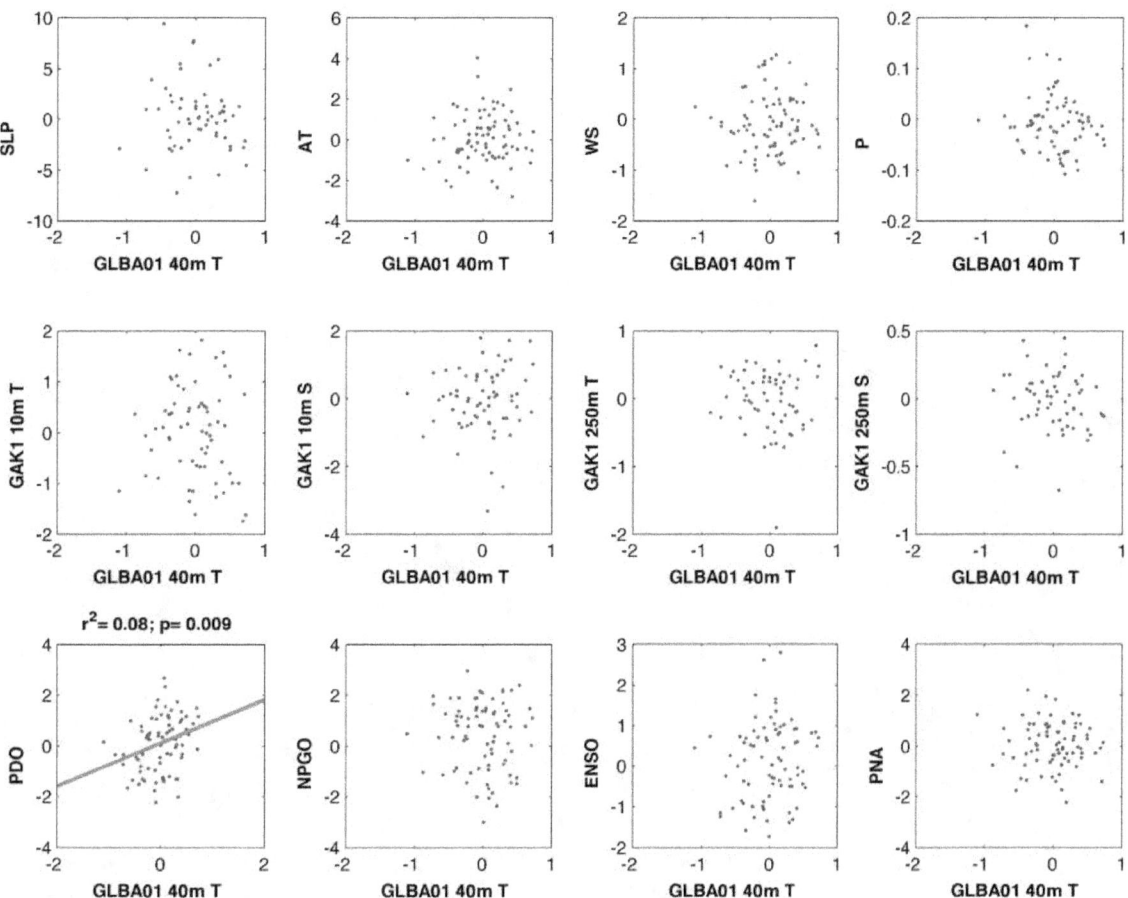

Glacier Bay Oceanographic Monitoring Program Analysis of Observations, 1993–2009
Appendix E. Regressions of temperature and salinity anomalies with atmospheric, oceanographic and climate time series.

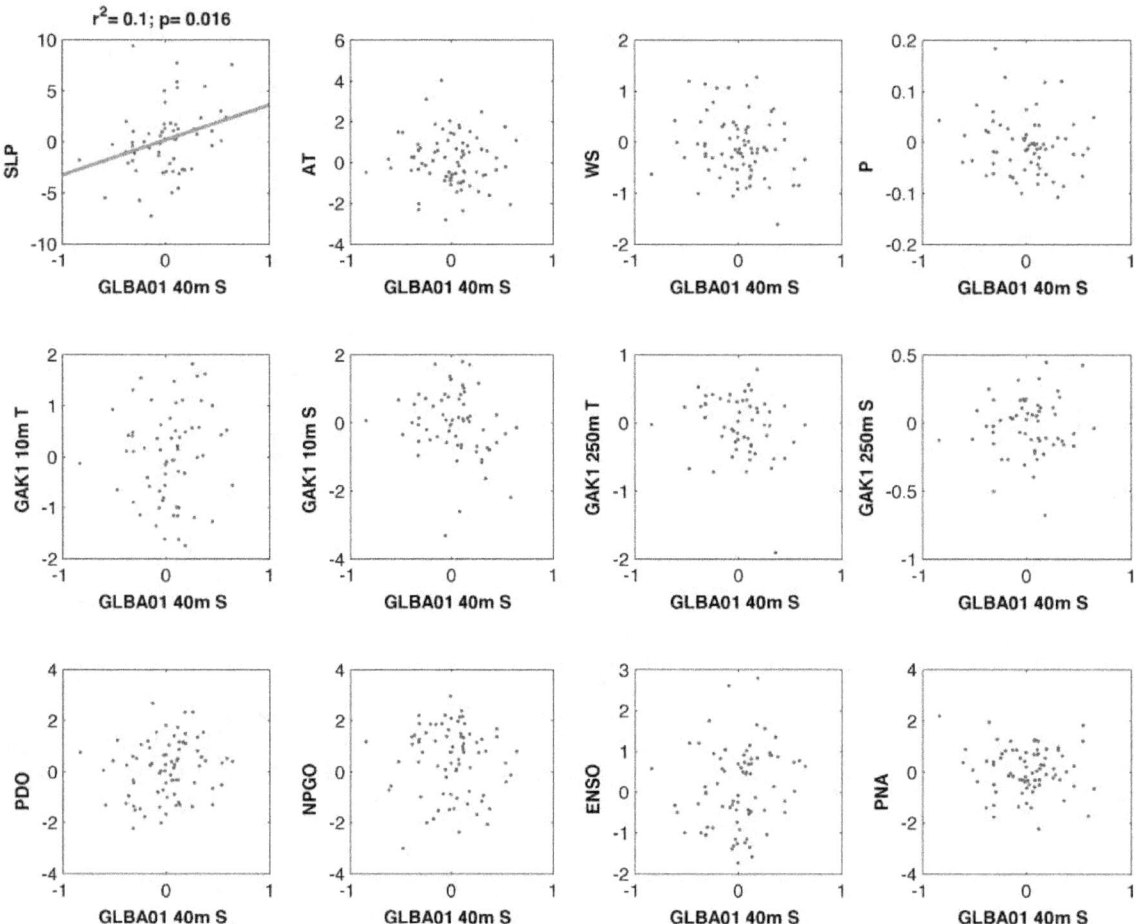

Glacier Bay Oceanographic Monitoring Program Analysis of Observations, 1993–2009
Appendix E. Regressions of temperature and salinity anomalies with atmospheric, oceanographic and climate time series.

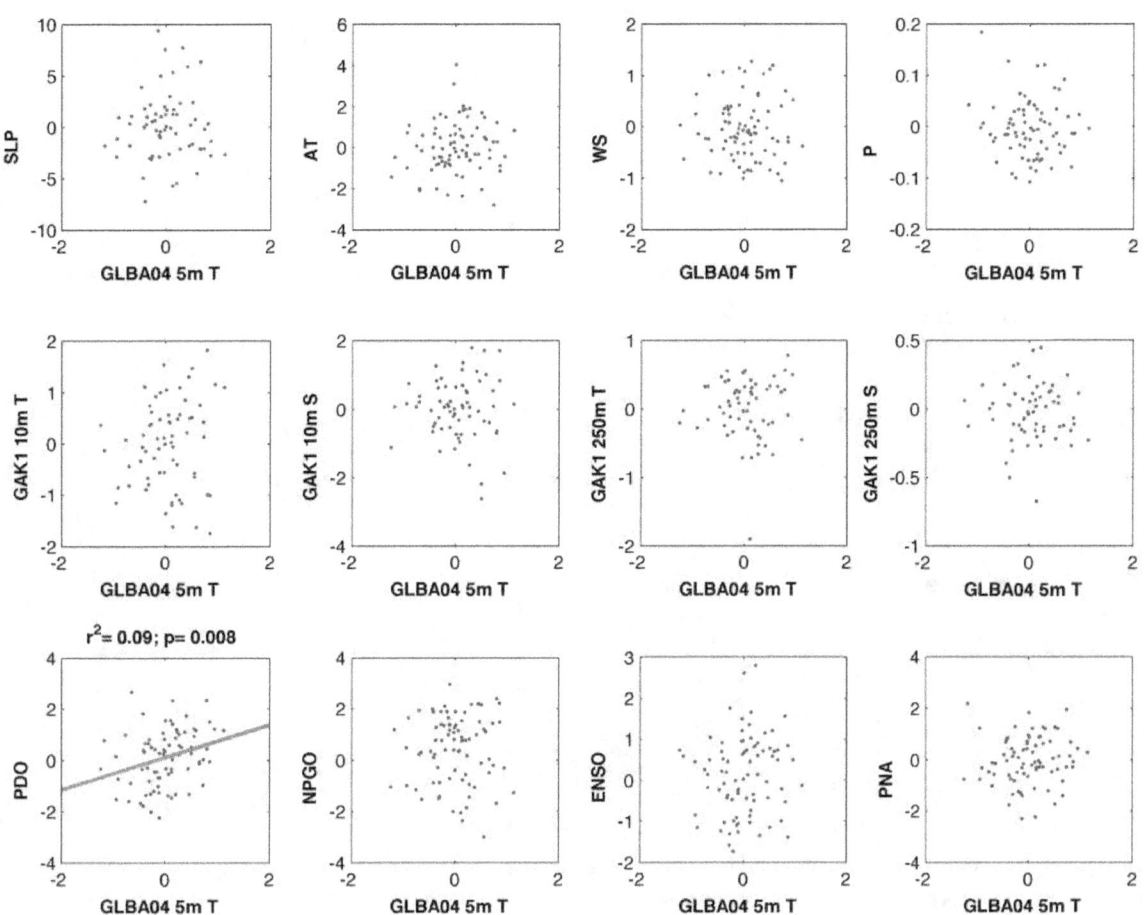

Glacier Bay Oceanographic Monitoring Program Analysis of Observations, 1993–2009
Appendix E. Regressions of temperature and salinity anomalies with atmospheric, oceanographic and climate time series.

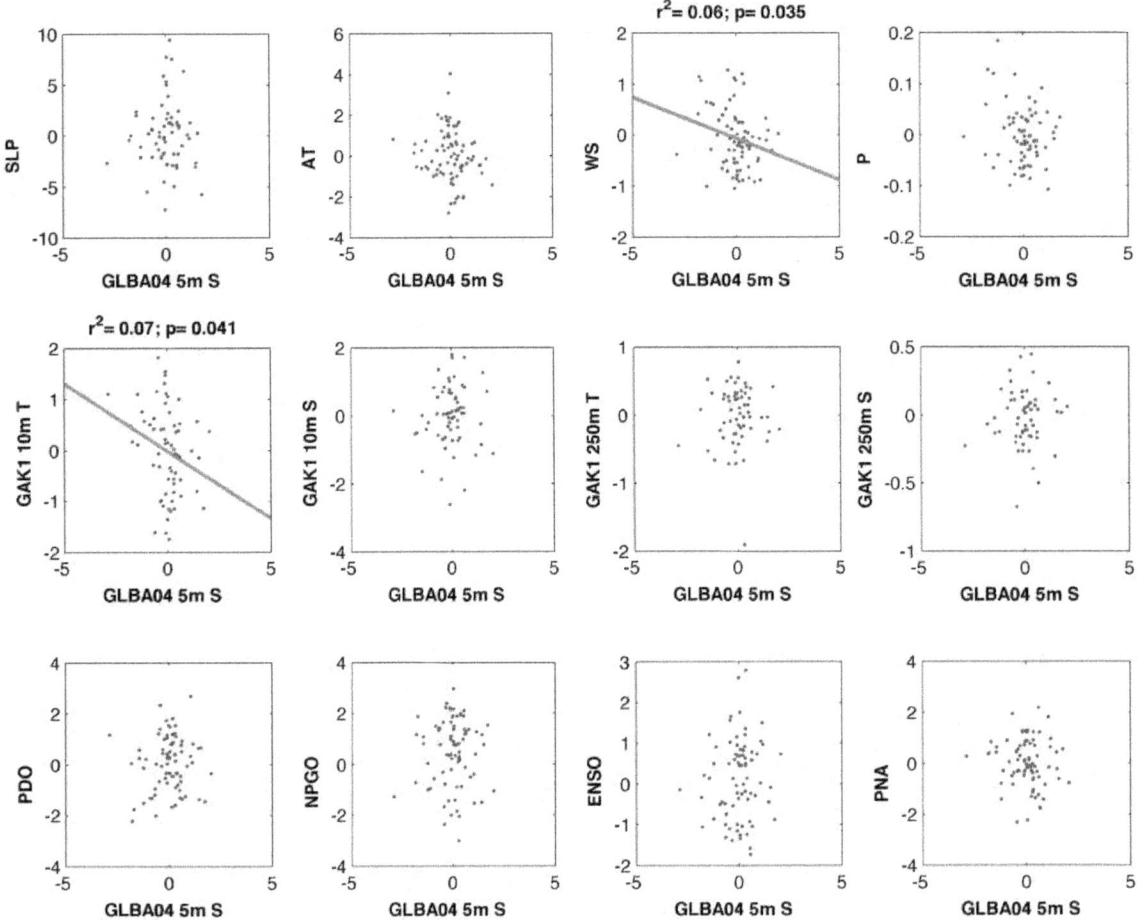

Glacier Bay Oceanographic Monitoring Program Analysis of Observations, 1993–2009
Appendix E. Regressions of temperature and salinity anomalies with atmospheric, oceanographic and climate time series.

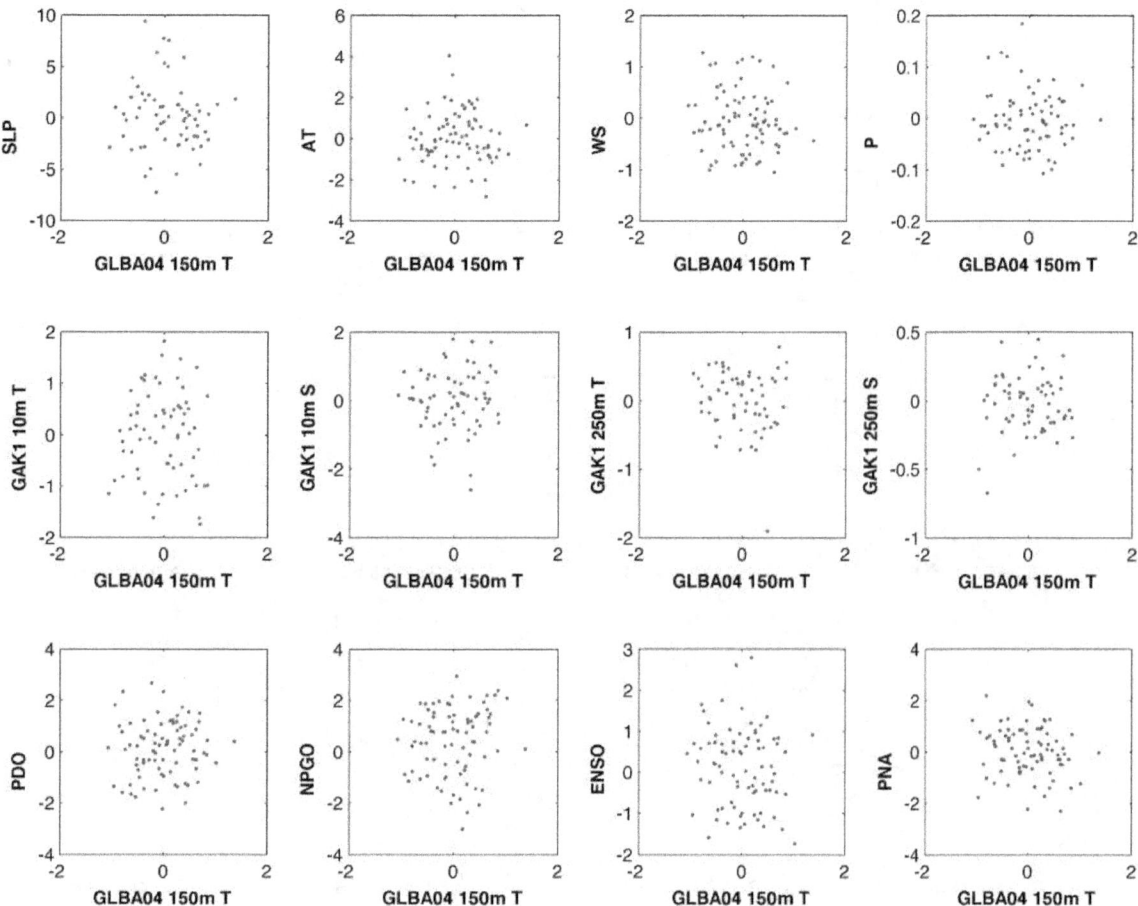

Glacier Bay Oceanographic Monitoring Program Analysis of Observations, 1993–2009
Appendix E. Regressions of temperature and salinity anomalies with atmospheric, oceanographic and climate time series.

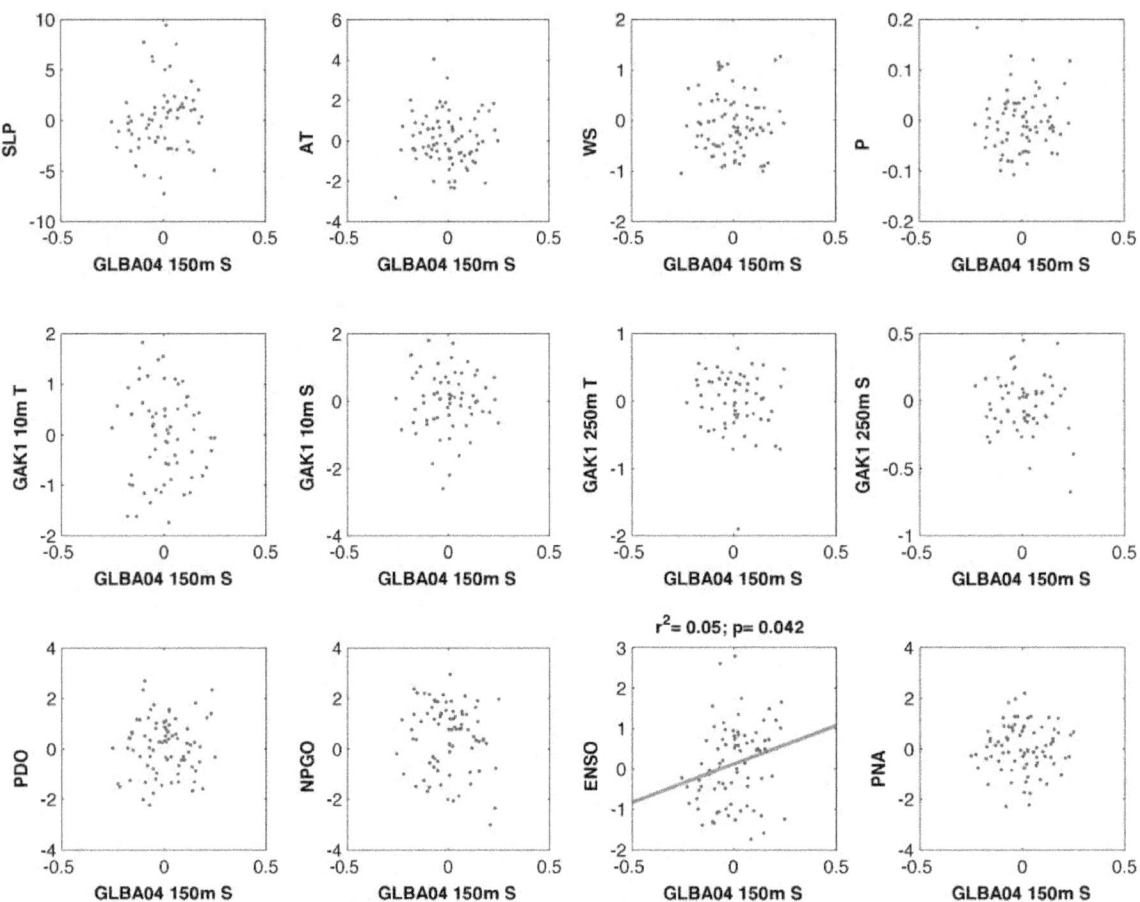

Glacier Bay Oceanographic Monitoring Program Analysis of Observations, 1993–2009
Appendix E. Regressions of temperature and salinity anomalies with atmospheric, oceanographic and climate time series.

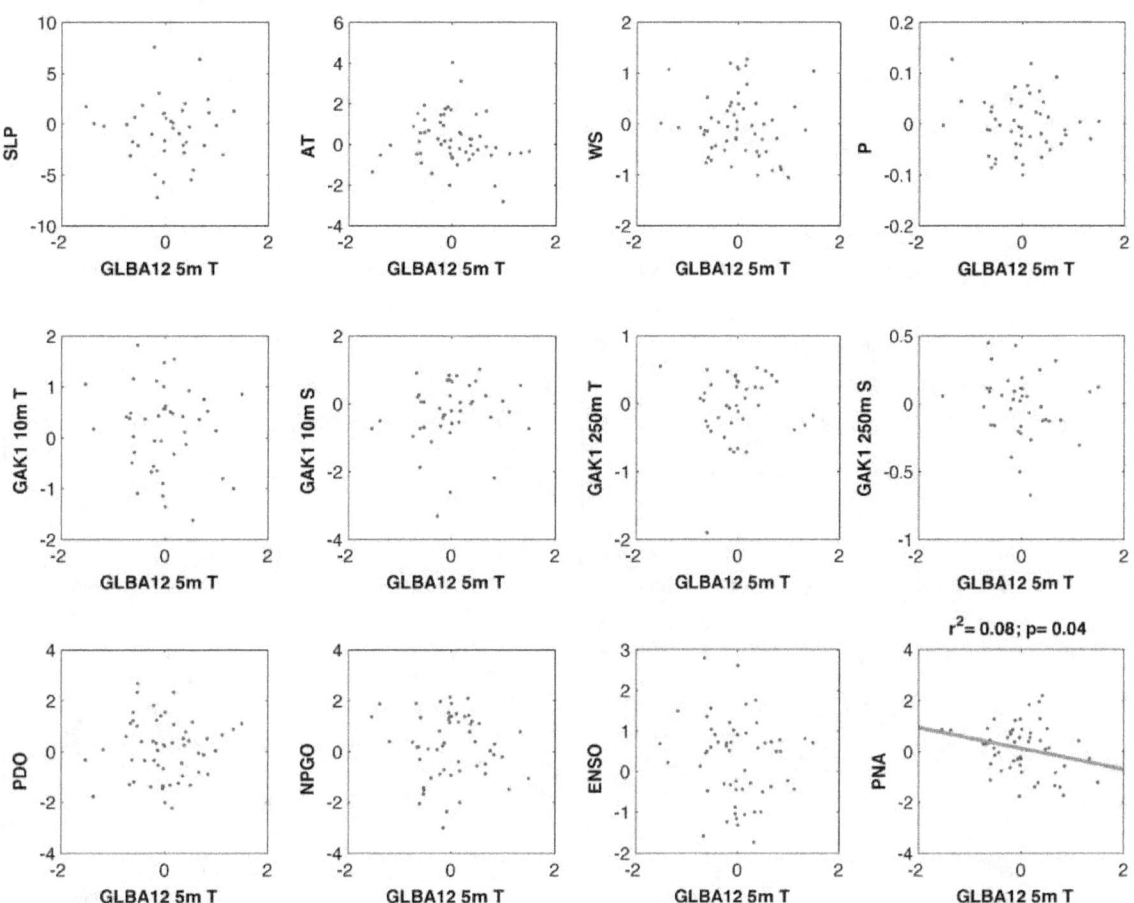

Glacier Bay Oceanographic Monitoring Program Analysis of Observations, 1993–2009
Appendix E. Regressions of temperature and salinity anomalies with atmospheric, oceanographic and climate time series.

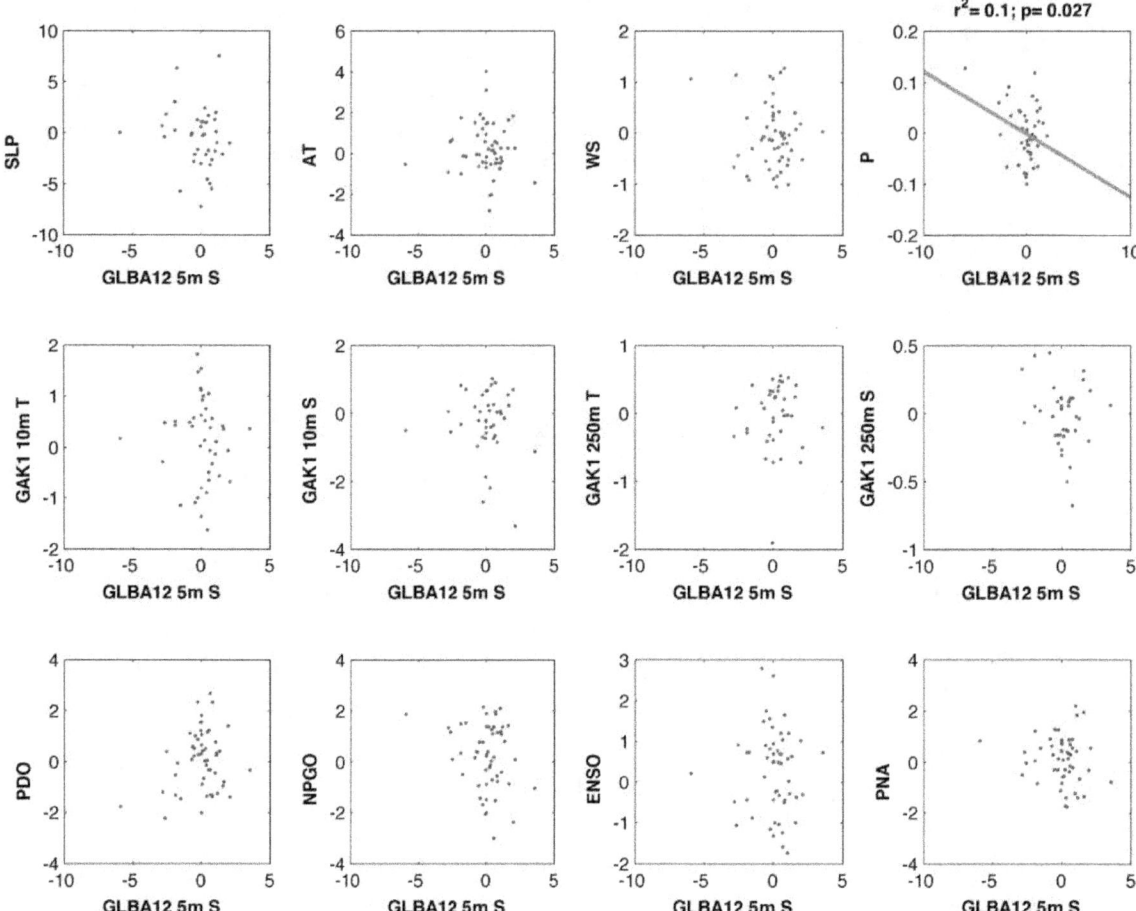

Glacier Bay Oceanographic Monitoring Program Analysis of Observations, 1993–2009
Appendix E. Regressions of temperature and salinity anomalies with atmospheric, oceanographic and climate time series.

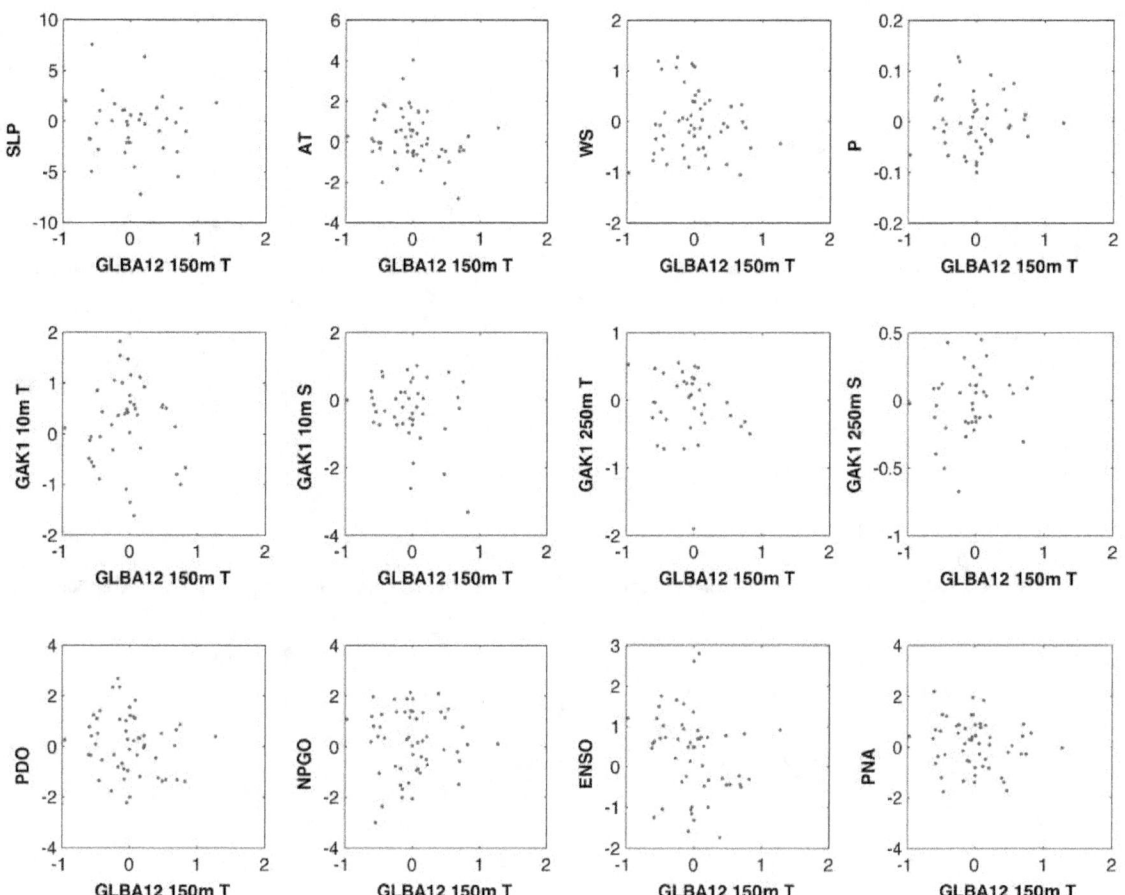

Glacier Bay Oceanographic Monitoring Program Analysis of Observations, 1993–2009
Appendix E. Regressions of temperature and salinity anomalies with atmospheric, oceanographic and climate time series.

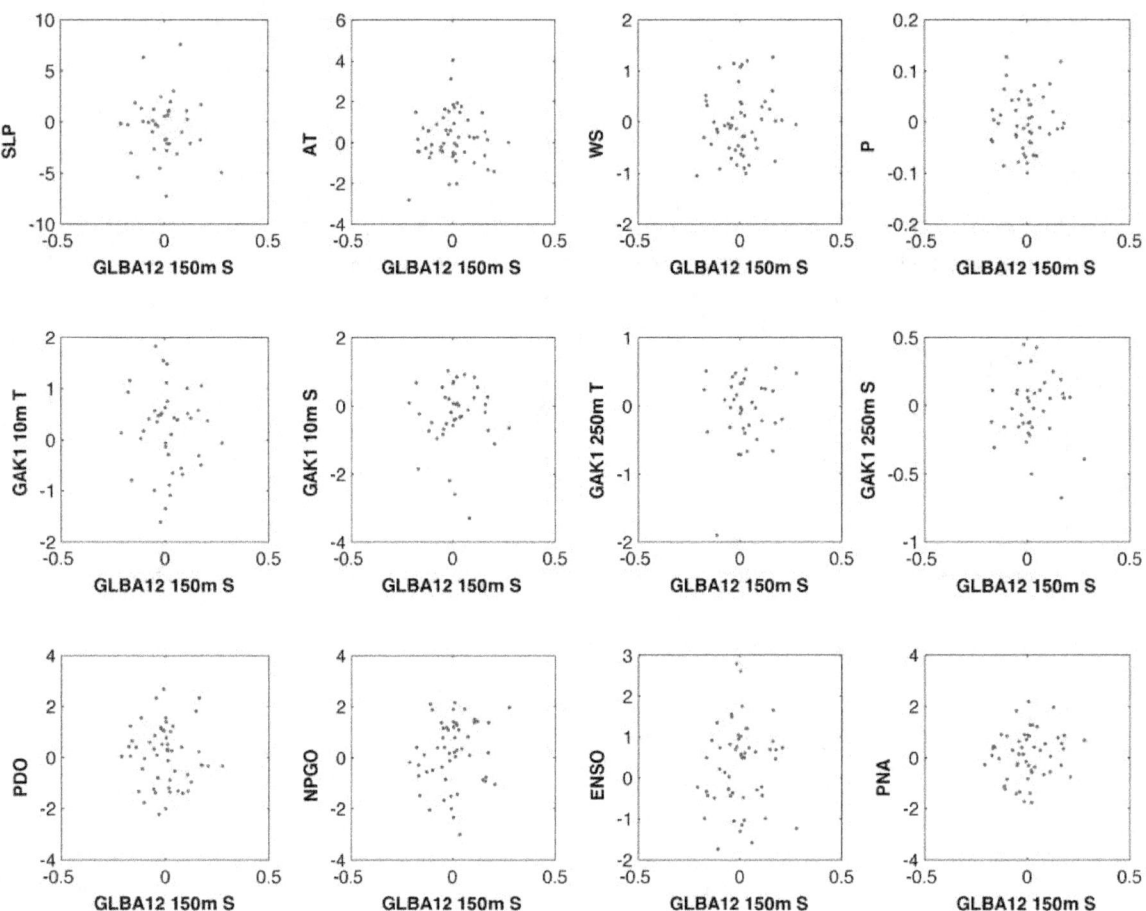

Glacier Bay Oceanographic Monitoring Program Analysis of Observations, 1993–2009
Appendix E. Regressions of temperature and salinity anomalies with atmospheric, oceanographic and climate time series.

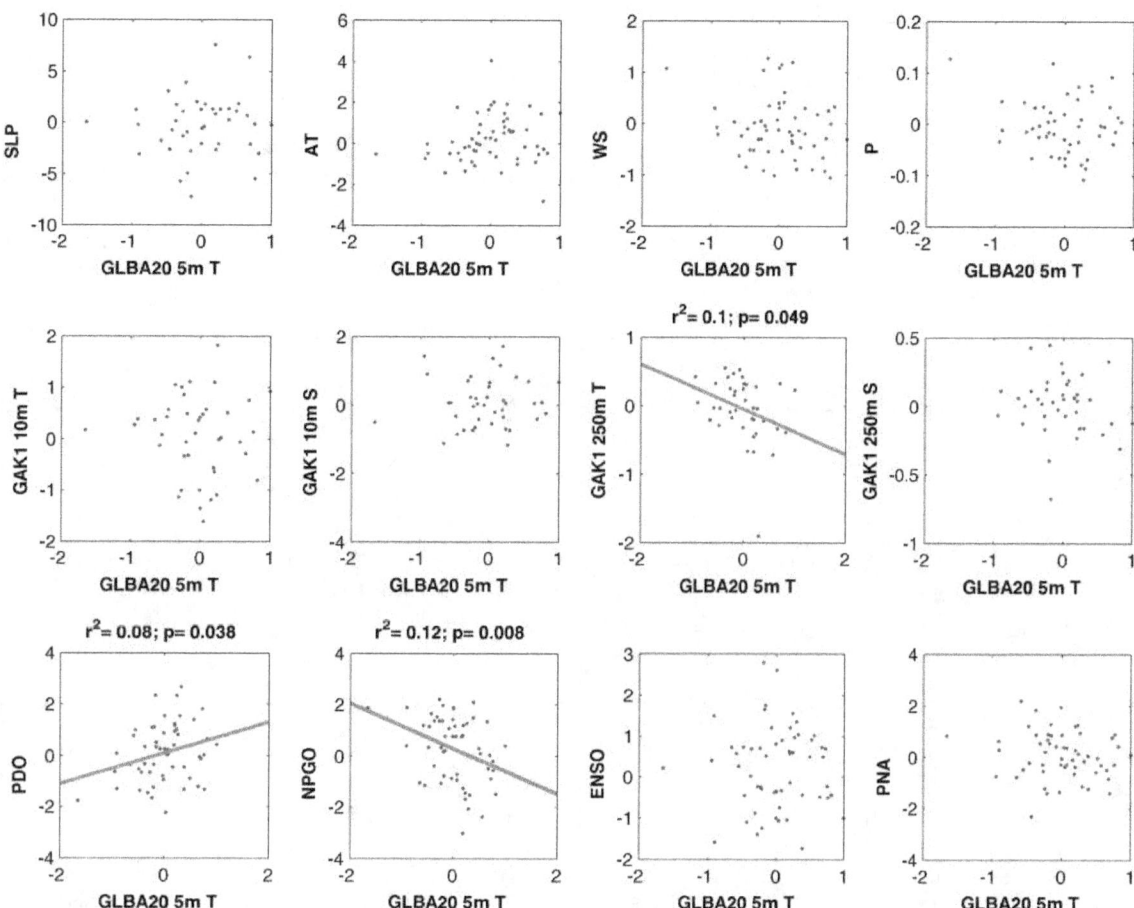

Glacier Bay Oceanographic Monitoring Program Analysis of Observations, 1993–2009
Appendix E. Regressions of temperature and salinity anomalies with atmospheric, oceanographic and climate time series.

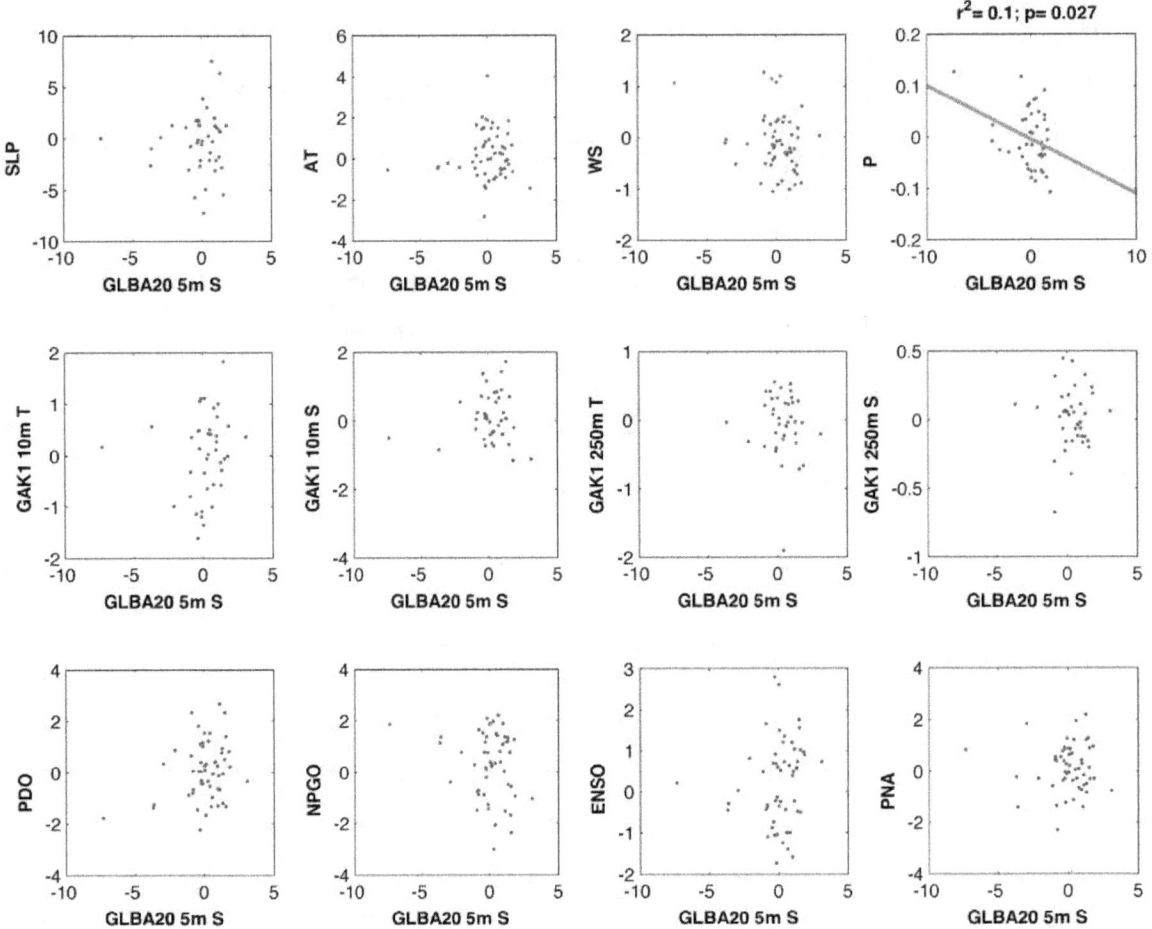

Glacier Bay Oceanographic Monitoring Program Analysis of Observations, 1993–2009
Appendix E. Regressions of temperature and salinity anomalies with atmospheric, oceanographic and climate time series.

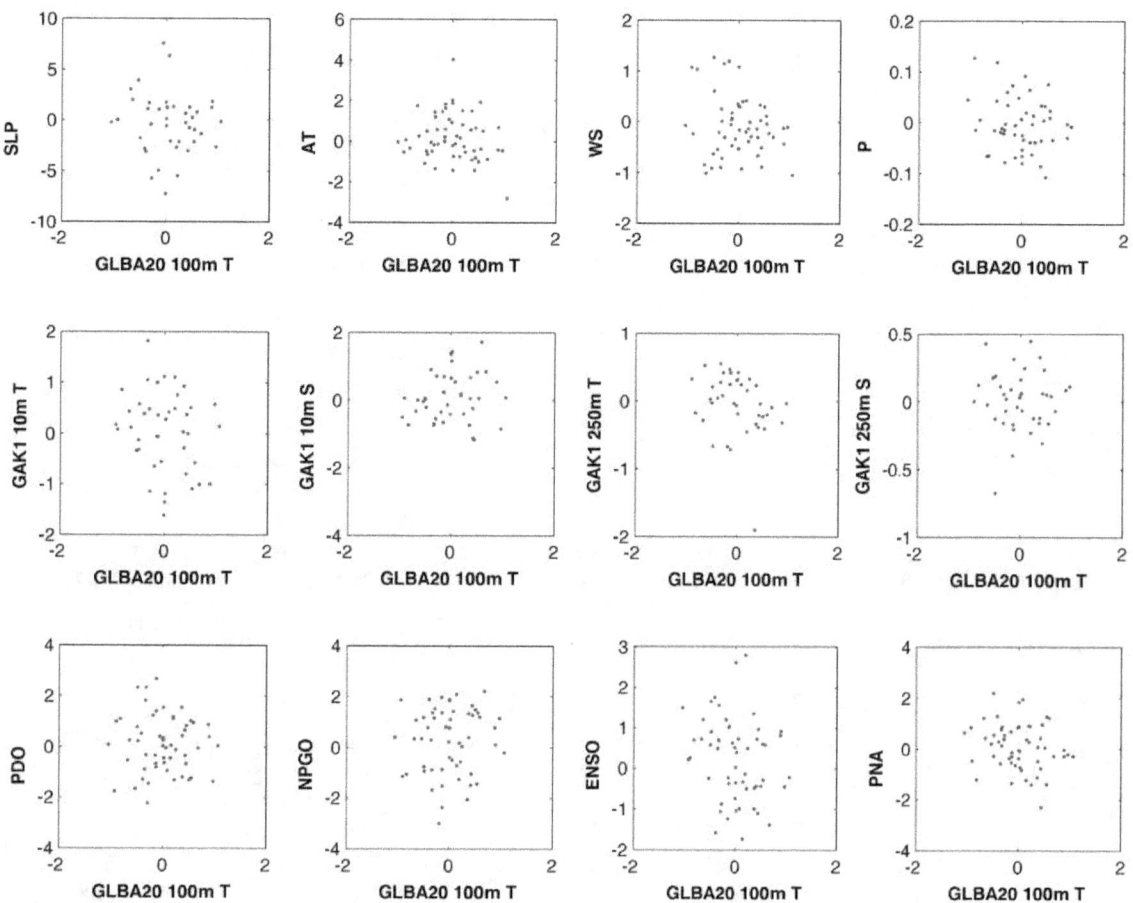

Glacier Bay Oceanographic Monitoring Program Analysis of Observations, 1993–2009
Appendix E. Regressions of temperature and salinity anomalies with atmospheric, oceanographic and climate time series.

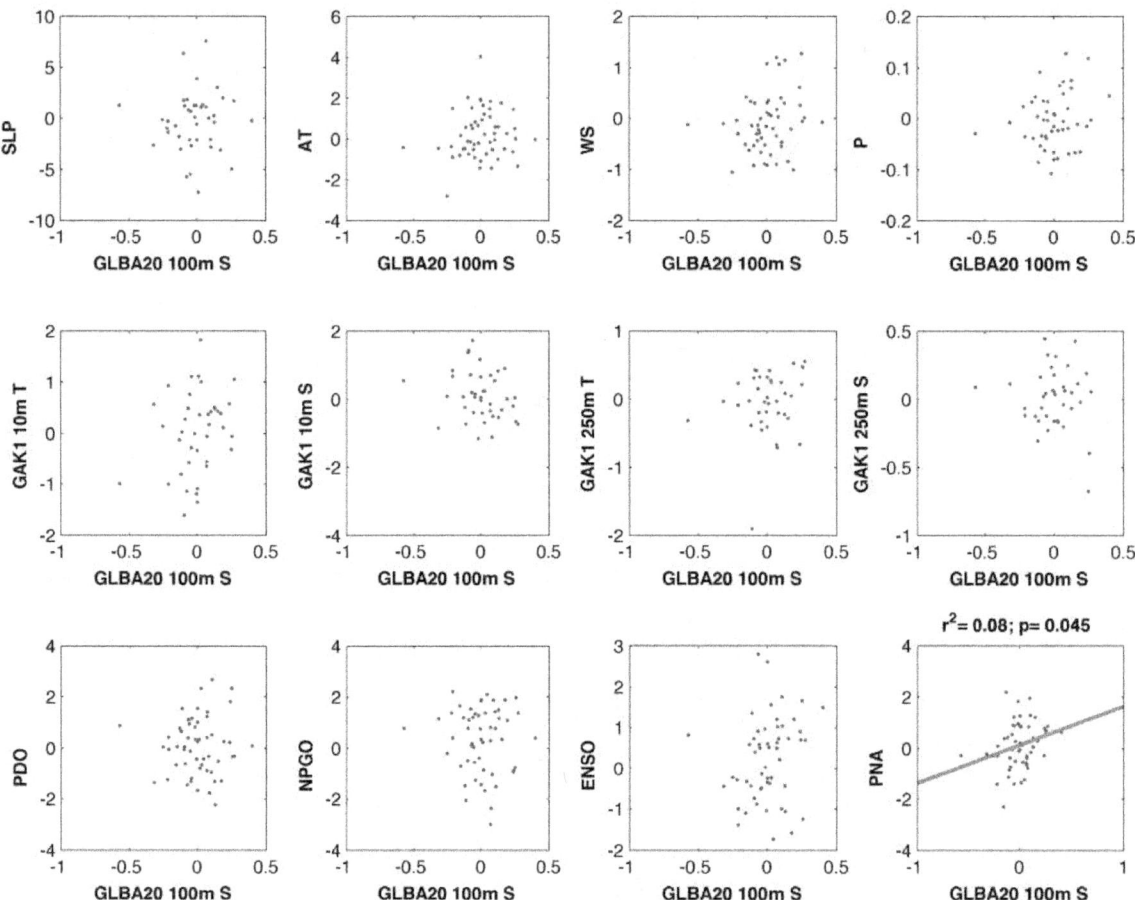

Appendix F. Tables of temperature and salinity correlations at each station between select depth levels.

```
STN DEPTH  0     10    20    30    50    75    100   150   200   250
00    0   1.00
00   10   0.47  1.00
00   20   0.18  0.63  1.00
00   30   ---   0.33  0.80  1.00
00   50   ---   0.34  0.26  ---   1.00
00   75   ...   ...   ...   ...   ...   ...
00  100   ...   ...   ...   ...   ...   ...   ...
00  150   ...   ...   ...   ...   ...   ...   ...   ...
00  200   ...   ...   ...   ...   ...   ...   ...   ...   ...
00  250   ...   ...   ...   ...   ...   ...   ...   ...   ...   ...
01    0   1.00
01   10   0.72  1.00
01   20   0.45  0.78  1.00
01   30   ---   0.21  0.56  1.00
01   50   ---   ---   0.12  0.53  1.00
01   75   ...   ...   ...   ...   ...   ...
01  100   ...   ...   ...   ...   ...   ...   ...
01  150   ...   ...   ...   ...   ...   ...   ...   ...
01  200   ...   ...   ...   ...   ...   ...   ...   ...   ...
01  250   ...   ...   ...   ...   ...   ...   ...   ...   ...   ...
02    0   1.00
02   10   0.64  1.00
02   20   0.29  0.78  1.00
02   30   0.12  0.53  0.85  1.00
02   50   0.06  0.31  0.61  0.84  1.00
02   75   ---   0.12  0.29  0.45  0.68  1.00
02  100   ...   ...   ...   ...   ...   ...   ...
02  150   ...   ...   ...   ...   ...   ...   ...   ...
02  200   ...   ...   ...   ...   ...   ...   ...   ...   ...
02  250   ...   ...   ...   ...   ...   ...   ...   ...   ...   ...
03    0   1.00
03   10   0.29  1.00
03   20   0.20  0.73  1.00
03   30   0.06  0.49  0.74  1.00
03   50   ---   0.33  0.57  0.86  1.00
03   75   ---   0.20  0.46  0.73  0.88  1.00
03  100   ---   ---   ---   0.43  0.75  0.82  1.00
03  150   ...   ...   ...   ...   ...   ...   ...   ...
03  200   ...   ...   ...   ...   ...   ...   ...   ...   ...
03  250   ...   ...   ...   ...   ...   ...   ...   ...   ...   ...
04    0   1.00
04   10   ---   1.00
04   20   ---   0.43  1.00
04   30   ---   0.20  0.69  1.00
04   50   ---   0.12  0.50  0.78  1.00
04   75   ---   0.13  0.45  0.67  0.92  1.00
04  100   ---   0.10  0.45  0.63  0.88  0.96  1.00
04  150   ---   0.07  0.34  0.49  0.72  0.81  0.86  1.00
04  200   ---   0.08  0.28  0.39  0.57  0.64  0.66  0.85  1.00
04  250   ---   ---   0.22  0.29  0.37  0.44  0.44  0.67  0.88  1.00
05    0   1.00
05   10   ---   1.00
05   20   ---   0.42  1.00
05   30   ---   0.34  0.83  1.00
05   50   ---   0.25  0.61  0.74  1.00
05   75   ---   0.17  0.55  0.69  0.96  1.00
05  100   ---   0.15  0.51  0.65  0.94  0.98  1.00
05  150   ---   0.14  0.39  0.53  0.79  0.85  0.88  1.00
05  200   ---   0.15  0.27  0.39  0.60  0.67  0.69  0.90  1.00
05  250   ---   0.17  0.25  0.33  0.51  0.56  0.57  0.79  0.93  1.00
06    0   1.00
06   10   ---   1.00
06   20   ---   0.67  1.00
06   30   ---   0.54  0.89  1.00
06   50   ---   0.39  0.77  0.89  1.00
06   75   ---   0.29  0.64  0.77  0.94  1.00
06  100   ---   0.24  0.57  0.71  0.87  0.95  1.00
06  150   ---   0.19  0.42  0.55  0.69  0.78  0.89  1.00
06  200   ---   0.13  0.29  0.39  0.49  0.59  0.72  0.90  1.00
06  250   ---   0.11  0.18  0.26  0.33  0.43  0.54  0.70  0.87  1.00
07    0   1.00
07   10   0.13  1.00
```

Glacier Bay Oceanographic Monitoring Program Analysis of Observations, 1993–2009
Appendix F. Tables of temperature and salinity correlations at each station between select depth levels.

```
07   20  ---  0.55 1.00
07   30  ---  0.52 0.86 1.00
07   50  ---  0.43 0.82 0.92 1.00
07   75  ---  0.41 0.76 0.87 0.97 1.00
07  100  ---  0.36 0.72 0.81 0.92 0.95 1.00
07  150  ---  0.30 0.57 0.65 0.75 0.81 0.88 1.00
07  200  ---  0.21 0.29 0.37 0.47 0.54 0.64 0.86 1.00
07  250  ---  0.22 0.29 0.32 0.44 0.51 0.61 0.82 0.96 1.00
08    0  1.00
08   10  0.09 1.00
08   20  ---  0.41 1.00
08   30  ---  0.36 0.82 1.00
08   50  ---  0.30 0.75 0.91 1.00
08   75  ---  0.27 0.67 0.82 0.95 1.00
08  100  ---  0.24 0.62 0.78 0.92 0.97 1.00
08  150  0.08 0.15 0.46 0.62 0.76 0.86 0.90 1.00
08  200  0.09 ---  0.26 0.42 0.52 0.60 0.65 0.87 1.00
08  250  0.10 0.12 0.26 0.38 0.48 0.53 0.55 0.73 0.91 1.00


STN DEPTH   0    10   20   30   50   75   100  150  200  250
09    0  1.00
09   10  0.10 1.00
09   20  ---  0.31 1.00
09   30  ---  0.27 0.92 1.00
09   50  ---  0.30 0.65 0.85 1.00
09   75  ---  0.29 0.63 0.80 0.91 1.00
09  100  ---  0.29 0.57 0.71 0.82 0.95 1.00
09  150  ---  0.13 0.42 0.56 0.67 0.85 0.89 1.00
09  200  ---  0.08 0.24 0.37 0.50 0.67 0.73 0.90 1.00
09  250  ---  ---  0.20 0.32 0.43 0.57 0.63 0.78 0.94 1.00
10    0  1.00
10   10  0.08 1.00
10   20  ---  0.43 1.00
10   30  ---  0.28 0.86 1.00
10   50  ---  0.24 0.67 0.89 1.00
10   75  ---  0.23 0.60 0.79 0.92 1.00
10  100  ---  0.24 0.55 0.70 0.84 0.95 1.00
10  150  ---  0.10 0.32 0.47 0.66 0.78 0.87 1.00
10  200  0.10 ---  0.18 0.26 0.43 0.57 0.68 0.90 1.00
10  250  0.08 0.10 0.14 0.21 0.33 0.47 0.56 0.74 0.89 1.00
11    0  1.00
11   10  0.15 1.00
11   20  ---  0.77 1.00
11   30  ---  0.41 0.70 1.00
11   50  ---  0.29 0.51 0.85 1.00
11   75  ---  0.20 0.38 0.66 0.87 1.00
11  100  ---  0.14 0.27 0.54 0.77 0.90 1.00
11  150  ---  ---  0.16 0.40 0.64 0.78 0.90 1.00
11  200  ---  ---  ---  0.24 0.41 0.55 0.71 0.86 1.00
11  250  ---  ---  ---  0.19 0.32 0.43 0.61 0.78 0.94 1.00
12    0  1.00
12   10  0.25 1.00
12   20  ---  0.47 1.00
12   30  ---  0.32 0.77 1.00
12   50  0.17 0.25 0.56 0.80 1.00
12   75  0.10 0.25 0.49 0.63 0.71 1.00
12  100  0.14 0.20 0.38 0.49 0.63 0.88 1.00
12  150  0.17 0.14 0.29 0.37 0.61 0.71 0.85 1.00
12  200  ---  0.09 0.20 0.17 0.29 0.40 0.54 0.74 1.00
12  250  0.12 0.15 0.25 0.23 0.32 0.39 0.54 0.71 0.85 1.00
13    0  1.00
13   10  ---  1.00
13   20  ---  0.53 1.00
13   30  ---  0.43 0.88 1.00
13   50  ---  0.38 0.76 0.82 1.00
13   75  ---  0.33 0.62 0.67 0.89 1.00
13  100  ---  0.25 0.52 0.57 0.79 0.93 1.00
13  150  ...  ...  ...  ...  ...  ...  ...
13  200  ...  ...  ...  ...  ...  ...  ...
13  250  ...  ...  ...  ...  ...  ...  ...  ...  ...
```

Glacier Bay Oceanographic Monitoring Program Analysis of Observations, 1993–2009
Appendix F. Tables of temperature and salinity correlations at each station between select depth levels.

```
14    0  1.00
14   10  --- 1.00
14   20  --- 0.61 1.00
14   30  0.08 0.42 0.78 1.00
14   50  0.09 0.27 0.55 0.73 1.00
14   75  --- --- --- --- --- 1.00
14  100  ... ... ... ... ... ... ...
14  150  ... ... ... ... ... ... ... ...
14  200  ... ... ... ... ... ... ... ... ...
14  250  ... ... ... ... ... ... ... ... ... ...
15    0  1.00
15   10  --- 1.00
15   20  --- 0.61 1.00
15   30  --- 0.62 0.95 1.00
15   50  --- --- --- --- 1.00
15   75  --- --- --- --- 0.79 1.00
15  100  --- --- --- --- 0.27 0.58 1.00
15  150  ... ... ... ... ... ... ... ...
15  200  ... ... ... ... ... ... ... ... ...
15  250  ... ... ... ... ... ... ... ... ... ...
16    0  1.00
16   10  --- 1.00
16   20  --- 0.54 1.00
16   30  --- 0.42 0.86 1.00
16   50  --- 0.37 0.74 0.91 1 00
16   75  --- 0.37 0.65 0.81 0 93 1.00
16  100  --- 0.24 0.42 0.57 0.72 0.88 1.00
16  150  0.09 0.07 0.15 0.27 0.42 0.58 0.84 1.00
16  200  0.11 0.08 0.16 0.27 0.44 0.59 0.81 0.97 1.00
16  250  0.15 0.07 0.16 0.27 0.43 0.57 0.79 0.95 0.99 1.00
17    0  1.00
17   10  --- 1.00
17   20  --- 0.55 1.00
17   30  --- 0.50 0.84 1.00
17   50  --- 0.50 0.65 0.86 1.00
17   75  --- 0.30 0.45 0.69 0.91 1.00
17  100  --- 0.20 0.30 0.56 0.78 0.93 1 00
17  150  --- --- 0.10 0.31 0.45 0.64 0.82 1.00
17  200  0.00 0.00 -0 00 0.00 0.00 0.00 0.00 0.00 ...
17  250  ... ... ... ... ... ... ... ... ... ...
STN DEPTH  0   10   20   30   50   75  100  150  200  250
18    0  1.00
18   10  --- 1.00
18   20  --- 0.36 1 00
18   30  --- 0.32 0 89 1.00
18   50  --- 0.29 0.68 0.85 1.00
18   75  --- 0.25 0.57 0.74 0.92 1.00
18  100  --- 0.10 0 33 0.47 0.66 0.84 1.00
18  150  --- --- 0.14 0.25 0.42 0.59 0.82 1.00
18  200  --- --- --- --- --- --- --- --- 1.00
18  250  ... ... ... ... ... ... ... ... ...
19    0  1.00
19   10  --- 1.00
19   20  --- 0.68 1.00
19   30  --- 0.64 0.84 1.00
19   50  --- 0.36 0.45 0.73 1.00
19   75  --- 0.26 0.37 0.61 0.93 1.00
19  100  --- 0.14 0.22 0.40 0.76 0.89 1.00
19  150  --- --- --- 0.20 0.48 0.62 0.79 1.00
19  200  --- --- --- 0.19 0.45 0.54 0.68 0.94 1.00
19  250  ... ... ... ... ... ... ... ... ... ...
20    0  1.00
20   10  --- 1.00
20   20  --- 0.39 1.00
20   30  --- 0.41 0.70 1.00
20   50  --- 0.14 0.23 0.65 1.00
20   75  --- --- 0.15 0.47 0.87 1.00
20  100  --- --- 0.16 0.40 0.75 0.85 1.00
20  150  --- --- --- 0.26 0.60 0.66 0.79 1.00
20  200  ... ... ... ... ... ... ... ...
20  250  ... ... ... ... ... ... ... ... ...
21    0  1.00
21   10  --- 1.00
21   20  --- 0.25 1.00
21   30  --- 0.31 0.43 1.00
21   50  --- 0.23 0.19 0.61 1 00
21   75  --- 0.17 0.17 0.43 0.76 1.00
21  100  --- 0.16 0.15 0.46 0.88 0.92 1.00
21  150  --- --- --- 0.18 0.53 0.83 0.78 1.00
21  200  --- --- --- --- --- --- --- --- 1.00
```

Glacier Bay Oceanographic Monitoring Program Analysis of Observations, 1993–2009
Appendix F. Tables of temperature and salinity correlations at each station between select depth levels.

```
21   250   ---  ---  ---  ---  ---  ---  ---  ---  ---
22     0  1.00
22    10  0.17 1.00
22    20  0.20 0.70 1.00
22    30  0.13 0.51 0.83 1.00
22    50   ---  --- 0.32 0.60 1.00
22    75   ---  --- 0.26 0.53 0.98 1.00
22   100   ---  --- 0.19 0.44 0.92 0.95 1.00
22   150   ---  ---  ---  ---  ---  --- 0.54 1.00
22   200   ---  ---  ---  ---  ---  ---  ---  ---
22   250   ---  ---  ---  ---  ---  ---  ---  ---  ---
23     0  1.00
23    10   --- 1.00
23    20   --- 0.19 1.00
23    30   ---  --- 0.85 1.00
23    50   ---  --- 0.59 0.82 1.00
23    75  0.00 0.00 0.00 0.00 0.00 ...
23   100   ---  ---  ---  ---  ---  ---
23   150   ---  ---  ---  ---  ---  ---  ---
23   200   ---  ---  ---  ---  ---  ---  ---  ---
23   250   ---  ---  ---  ---  ---  ---  ---  ---  ---
```

Glacier Bay Oceanographic Monitoring Program Analysis of Observations, 1993–2009
Appendix F. Tables of temperature and salinity correlations at each station between select depth levels.

Table of temperature correlations at each station between select depth levels.

STN	DEPTH	0	10	20	30	50	75	100	150	200	250
00	0	1.00									
00	10	0.67	1.00								
00	20	0.40	0.79	1.00							
00	30	0.16	0.62	0.84	1.00						
00	50	---	0.20	0.28	0.39	1.00					
00	75	...	...	...	...	...					
00	100	...	...	...	...	...	...				
00	150	...	...	...	...	...	...	...			
00	200	...	...	...	...	...	...	...	...		
00	250	...	...	...	...	...	...	...	...	...	
01	0	1.00									
01	10	0.79	1.00								
01	20	0.72	0.95	1.00							
01	30	0.60	0.81	0.94	1.00						
01	50	0.50	0.58	0.69	0.78	1.00					
01	75	...	...	...	...	...					
01	100	...	...	...	...	...	...				
01	150	...	...	...	...	...	...	...			
01	200	...	...	...	...	...	...	...	...		
01	250	...	...	...	...	...	...	...	...	...	
02	0	1.00									
02	10	0.62	1.00								
02	20	0.44	0.95	1.00							
02	30	0.37	0.88	0.95	1.00						
02	50	0.38	0.83	0.87	0.95	1.00					
02	75	0.39	0.67	0.70	0.78	0.87	1.00				
02	100	...	...	...	...	...	...				
02	150	...	...	...	...	...	...	...			
02	200	...	...	...	...	...	...	...	...		
02	250	...	...	...	...	...	...	...	...	...	
03	0	1.00									
03	10	0.50	1.00								
03	20	0.44	0.86	1.00							
03	30	0.37	0.75	0.96	1.00						
03	50	0.32	0.65	0.88	0.95	1.00					
03	75	0.31	0.48	0.72	0.80	0.87	1.00				
03	100	---	---	0.28	0.32	0.40	0.59	1.00			
03	150	...	...	...	...	...	...	...			
03	200	...	...	...	...	...	...	...	...		
03	250	...	...	...	...	...	...	...	...	...	
04	0	1.00									
04	10	0.25	1.00								
04	20	0.26	0.81	1.00							
04	30	0.17	0.65	0.84	1.00						
04	50	0.08	0.47	0.63	0.84	1.00					
04	75	0.06	0.35	0.49	0.68	0.90	1.00				
04	100	---	0.27	0.41	0.61	0.83	0.91	1.00			
04	150	---	0.18	0.31	0.48	0.64	0.74	0.87	1.00		
04	200	---	0.16	0.30	0.43	0.54	0.63	0.77	0.89	1.00	
04	250	---	0.12	0.23	0.35	0.38	0.47	0.63	0.78	0.90	1.00
05	0	1.00									
05	10	0.14	1.00								
05	20	0.10	0.70	1.00							
05	30	0.10	0.62	0.88	1.00						
05	50	---	0.58	0.83	0.85	1.00					
05	75	---	0.50	0.70	0.74	0.93	1.00				
05	100	---	0.49	0.66	0.67	0.87	0.90	1.00			
05	150	---	0.25	0.39	0.44	0.59	0.66	0.77	1.00		
05	200	---	0.17	0.27	0.27	0.37	0.42	0.56	0.84	1.00	
05	250	---	0.16	0.21	0.20	0.30	0.34	0.46	0.72	0.93	1.00
06	0	1.00									
06	10	0.09	1.00								
06	20	0.07	0.84	1.00							
06	30	0.06	0.78	0.95	1.00						
06	50	---	0.74	0.85	0.91	1.00					
06	75	---	0.68	0.76	0.81	0.89	1.00				
06	100	---	0.63	0.69	0.72	0.80	0.90	1.00			
06	150	---	0.51	0.49	0.52	0.58	0.61	0.78	1.00		
06	200	---	0.35	0.39	0.42	0.46	0.42	0.54	0.79	1.00	
06	250	---	0.40	0.38	0.35	0.36	0.24	0.33	0.62	0.85	1.00
07	0	1.00									
07	10	0.11	1.00								
07	20	---	0.63	1.00							

Glacier Bay Oceanographic Monitoring Program Analysis of Observations, 1993–2009
Appendix F. Tables of temperature and salinity correlations at each station between select depth levels.

```
07   30  0.07 0.49 0.90 1.00
07   50  ---  0.45 0.85 0.94 1.00
07   75  0.07 0.32 0.75 0.84 0.94 1.00
07  100  ---  0.27 0.68 0.76 0.85 0.92 1.00
07  150  ---  0.12 0.49 0.50 0.56 0.64 0.78 1.00
07  200  ---  ---  0.32 0.32 0.36 0.41 0.49 0.80 1.00
07  250  ---  ---  0.26 0.25 0.27 0.31 0.38 0.71 0.94 1.00
08    0  1.00
08   10  0.16 1.00
08   20  0.13 0.73 1.00
08   30  0.09 0.58 0.86 1.00
08   50  ---  0.57 0.80 0.87 1.00
08   75  ---  0.50 0.72 0.78 0.95 1.00
08  100  ---  0.43 0.66 0.78 0.91 0.93 1.00
08  150  ---  0.22 0.41 0.55 0.66 0.73 0.79 1.00
08  200  0.09 ---  0.20 0.34 0.38 0.43 0.52 0.76 1.00
08  250  ---  ---  0.17 0.32 0.34 0.38 0.44 0.67 0.95 1.00
```

```
STN DEPTH   0    10   20   30   50   75  100  150  200  250
09    0   1.00
09   10   0.09 1.00
09   20   0.12 0.55 1.00
09   30   0.12 0.58 0.88 1.00
09   50   ---  0.51 0.77 0.91 1.00
09   75   ---  0.52 0.74 0.87 0.91 1.00
09  100   0.07 0.52 0.71 0.81 0.84 0.93 1.00
09  150   ---  0.44 0.46 0.52 0.58 0.66 0.75 1.00
09  200   0.11 0.20 0.24 0.30 0.33 0.40 0.43 0.75 1.00
09  250   0.12 0.17 0.24 0.27 0.27 0.34 0.37 0.69 0.93 1.00
10    0   1.00
10   10   ---  1.00
10   20   ---  0.49 1.00
10   30   ---  0.44 0.94 1.00
10   50   0.07 0.38 0.82 0.91 1.00
10   75   ---  0.30 0.68 0.78 0.91 1.00
10  100   ---  0.37 0.70 0.76 0.87 0.94 1.00
10  150   0.08 0.22 0.56 0.61 0.73 0.81 0.86 1.00
10  200   0.08 0.08 0.40 0.42 0.55 0.59 0.60 0.79 1.00
10  250   0.15 ---  0.35 0.34 0.43 0.45 0.45 0.66 0.87 1.00
11    0   1.00
11   10   ---  1.00
11   20   ---  0.59 1.00
11   30   ---  0.44 0.81 1.00
11   50   ---  0.35 0.68 0.84 1.00
11   75   ---  0.21 0.48 0.66 0.85 1.00
11  100   ---  0.22 0.44 0.63 0.79 0.88 1.00
11  150   ---  0.18 0.38 0.48 0.64 0.71 0.76 1.00
11  200   ---  ---  0.20 0.25 0.40 0.46 0.48 0.84 1.00
11  250   ---  ---  0.17 0.21 0.33 0.37 0.39 0.77 0.93 1.00
12    0   1.00
12   10   0.48 1.00
12   20   0.30 0.78 1.00
12   30   0.23 0.63 0.88 1.00
12   50   0.17 0.51 0.74 0.91 1.00
12   75   0.11 0.34 0.50 0.69 0.83 1.00
12  100   0.13 0.31 0.44 0.62 0.76 0.96 1.00
12  150   0.13 0.28 0.39 0.52 0.62 0.79 0.82 1.00
12  200   0.10 0.10 0.16 0.27 0.34 0.58 0.63 0.86 1.00
12  250   0.10 0.11 0.15 0.21 0.25 0.43 0.42 0.68 0.82 1.00
13    0   1.00
13   10   0.24 1.00
13   20   0.15 0.84 1.00
13   30   0.11 0.76 0.92 1.00
13   50   0.11 0.63 0.75 0.81 1.00
13   75   0.13 0.48 0.57 0.61 0.86 1.00
13  100   0.15 0.38 0.43 0.46 0.72 0.89 1.00
13  150   ...  ...  ...  ...  ...  ...  ...
13  200   ...  ...  ...  ...  ...  ...  ...  ...
13  250   ...  ...  ...  ...  ...  ...  ...  ...  ...
14    0   1.00
14   10   0.16 1.00
```

Glacier Bay Oceanographic Monitoring Program Analysis of Observations, 1993–2009
Appendix F. Tables of temperature and salinity correlations at each station between select depth levels.

```
14   20  0.10 0.89 1.00
14   30  0.10 0.82 0.96 1.00
14   50  0.11 0.54 0.69 0.78 1.00
14   75  ---  ---  ---  ---  0.54 1.00
14   100 ...  ...  ...  ...  ...  ...
14   150 ...  ...  ...  ...  ...  ...  ...
14   200 ...  ...  ...  ...  ...  ...  ...  ...
14   250 ...  ...  ...  ...  ...  ...  ...  ...  ...
15    0  1.00
15   10  --- 1.00
15   20  --- 0.82 1.00
15   30  --- 0.72 0.92 1.00
15   50  --- 0.23 0.28 0.49 1.00
15   75  0.22 ---  ---  0.22 0.73 1.00
15   100 ---  ---  ---  ---  0.78 0.76 1.00
15   150 ...  ...  ...  ...  ...  ...  ...
15   200 ...  ...  ...  ...  ...  ...  ...  ...
15   250 ...  ...  ...  ...  ...  ...  ...  ...  ...
16    0  1.00
16   10  --- 1.00
16   20  --- 0.79 1.00
16   30  0.06 0.67 0.88 1.00
16   50  --- 0.53 0.64 0.79 1.00
16   75  --- 0.47 0.45 0.56 0.78 1.00
16   100 --- 0.37 0.34 0.45 0.65 0.91 1.00
16   150 0.15 0.30 0.26 0.38 0.53 0.70 0.86 1.00
16   200 0.19 0.27 0.23 0.33 0.45 0.62 0.77 0.96 1.00
16   250 0.19 0.28 0.23 0.32 0.44 0.63 0.76 0.94 0.99 1.00
17    0  1.00
17   10  --- 1.00
17   20  --- 0.78 1.00
17   30  --- 0.69 0.93 1.00
17   50  --- 0.64 0.72 0.83 1 00
17   75  --- 0.51 0.52 0.60 0 84 1.00
17   100 --- 0.44 0.45 0.54 0.73 0.88 1.00
17   150 0.07 0.36 0.34 0.40 0.49 0.64 0.84 1.00
17   200 0.00 0.00 0.00 0.00 0.00 0.00 -0 00 0.00 ...
17   250 ...  ...  ...  ...  ...  ...  ...  ...  ...
18    0  1.00

STN DEPTH  0   10   20   30   50   75  100  150  200  250
18   10  --- 1.00
18   20  --- 0.73 1.00
18   30  --- 0.64 0.94 1.00
18   50  --- 0.58 0.75 0.85 1.00
18   75  --- 0.59 0.63 0.72 0.89 1.00
18   100 0.07 0.57 0.53 0.56 0.67 0.86 1.00
18   150 0.07 0.50 0.47 0.48 0.54 0.70 0.88 1.00
18   200 --- 0.27 0.35 0.45 0.49 0.58 0.75 0.86 1.00
18   250 ...  ...  ...  ...  ...  ...  ...  ...
19    0  1.00
19   10  --- 1.00
19   20  --- 0.67 1 00
19   30  --- 0.55 0 93 1.00
19   50  --- 0.45 0 83 0.89 1.00
19   75  --- 0.42 0.75 0.80 0.92 1.00
19   100 --- 0.34 0.58 0.60 0.76 0.91 1 00
19   150 --- 0.30 0.56 0.54 0.69 0.80 0.89 1.00
19   200 --- 0.33 0.62 0.57 0.69 0.76 0.81 0.96 1 00
19   250 ...  ...  ...  ...  ...  ...  ...  ...  ...
20    0  1.00
20   10  --- 1.00
20   20  --- 0.71 1 00
20   30  --- 0.65 0 96 1.00
20   50  --- 0.57 0.85 0.94 1.00
20   75  0.09 0.55 0.74 0.84 0.94 1.00
20   100 --- 0.51 0.60 0.68 0.76 0.88 1.00
20   150 --- 0.45 0.58 0.65 0.67 0.75 0.81 1.00
20   200 ...  ...  ...  ...  ...  ...  ...  ...
20   250 ...  ...  ...  ...  ...  ...  ...  ...
21    0  1.00
21   10  0.45 1.00
21   20  0.15 0.72 1.00
21   30  0.17 0.68 0.90 1.00
21   50  0.18 0.59 0.81 0.93 1.00
21   75  0.16 0.54 0.79 0.88 0.95 1.00
21   100 --- 0.54 0.75 0.81 0.92 0.95 1.00
21   150 --- 0.46 0.55 0.64 0.75 0.82 0.91 1.00
21   200 0.64 ---  ---  ---  ---  ---  ---  ---  1.00
21   250 ...  ...  ...  ...  ...  ...  ...  ...
```

Glacier Bay Oceanographic Monitoring Program Analysis of Observations, 1993–2009
Appendix F. Tables of temperature and salinity correlations at each station between select depth levels.

```
22    0   1.00
22   10   0.17  1.00
22   20   0.18  0.89  1.00
22   30   0.16  0.62  0.77  1.00
22   50   ---   0.64  0.80  0.89  1.00
22   75   0.10  0.69  0.79  0.84  0.92  1.00
22  100   ---   0.64  0.78  0.80  0.89  0.95  1.00
22  150   ---   ---   ---   ---   0.44  0.63  0.76  1.00
22  200   ...   ...   ...   ...   ...   ...   ...   ...
22  250   ...   ...   ...   ...   ...   ...   ...   ...   ...
23    0   1.00
23   10   0.11  1.00
23   20   ---   0.88  1.00
23   30   ---   0.71  0.90  1.00
23   50   ---   0.67  0.84  0.94  1.00
23   75   0.00  -0.00 0.00  0.00  0.00  ...
23  100   ...   ...   ...   ...   ...   ...   ...
23  150   ...   ...   ...   ...   ...   ...   ...   ...
23  200   ...   ...   ...   ...   ...   ...   ...   ...   ...
23  250   ...   ...   ...   ...   ...   ...   ...   ...   ...
```

Appendix G. Tables of temperature and salinity correlations between stations at select depth levels.

Correlation of salinity at constant depth levels between station pairs; "—" denotes no significant relation found; "..." denotes no data taken at that depth level.

DEPTH STN	00	01	02	03	04	05	06	07	08	09	10	11	12	13	14	15	16	17	18	19	20	21	22	23
0 00	1.00																							
0 01	0.22	1.00																						
0 02	0.20	0.42	1.00																					
0 03	0.10	0.42	0.43	1.00																				
0 04	—	0.23	0.11	0.46	1.00																			
0 05	—	0.06	0.10	0.21	0.22	1.00																		
0 06	0.15	0.12	0.30	0.19	0.13	0.13	1.00																	
0 07	—	0.18	0.11	0.27	0.25	0.33	0.42	1.00																
0 08	—	—	—	0.13	0.19	0.31	0.22	0.47	1.00															
0 09	—	—	—	0.11	0.18	0.38	0.24	0.54	0.76	1.00														
0 10	—	—	—	0.11	0.19	0.41	0.38	0.53	0.55	0.49	1.00													
0 11	—	—	—	—	0.08	0.35	0.31	0.42	0.58	0.68	0.62	1.00												
0 12	—	—	0.09	0.10	0.09	0.35	0.41	0.44	0.47	0.53	0.60	0.61	1.00											
0 13	—	—	0.07	0.07	0.14	0.12	0.26	0.23	0.45	0.35	0.44	0.31	0.33	1.00										
0 14	—	—	0.10	0.16	0.14	0.17	0.14	0.34	0.52	0.39	0.47	0.34	0.34	0.30	0.49	1.00								
0 15	—	—	0.26	0.31	0.20	—	0.19	0.20	—	—	0.24	0.17	—	0.44	0.39	1.00								
0 16	—	—	—	—	—	0.07	0.08	0.30	0.28	0.35	0.37	0.25	0.36	0.66	0.35	1.00								
0 17	—	—	0.16	0.10	0.07	0.11	0.19	0.23	0.28	0.19	0.25	0.24	0.19	0.16	0.26	0.39	—	0.33	1.00					
0 18	—	—	0.12	0.13	0.22	0.17	0.32	0.16	0.27	0.22	0.15	0.28	0.18	0.10	0.27	0.29	0.23	0.17	0.15	1.00				
0 19	—	—	—	—	0.20	0.23	0.13	0.30	0.19	0.38	0.38	0.40	0.16	0.19	0.19	—	0.38	0.26	0.14	1.00				
0 20	—	—	—	—	—	—	—	0.13	—	0.17	—	0.12	—	0.28	—	0.23	0.24	—	0.20	1.00				
0 21	—	—	—	—	0.45	—	—	—	0.27	0.21	0.47	0.43	—	—	—	—	—	—	0.24	—	1.00			
0 22	—	0.22	—	0.22	0.29	0.38	—	0.40	0.25	0.27	0.19	—	0.17	—	—	—	—	0.14	0.22	—	—	1.00		
0 23	—	0.15	—	0.21	0.38	0.12	0.15	0.19	—	—	—	—	—	—	—	—	0.15	0.14	—	—	—	0.17	1.00	

DEPTH STN	00	01	02	03	04	05	06	07	08	09	10	11	12	13	14	15	16	17	18	19	20	21	22	23
10 00	1.00																							
10 01	0.17	1.00																						
10 02	0.11	0.38	1.00																					
10 03	—	0.15	0.30	1.00																				
10 04	0.23	0.22	0.28	0.33	1.00																			
10 05	0.10	—	0.10	0.18	0.24	1.00																		
10 06	—	0.13	—	0.39	0.11	0.33	1.00																	
10 07	0.07	—	0.36	0.08	0.14	0.60	1.00																	
10 08	—	—	—	0.13	—	0.13	0.60	0.66	1.00															
10 09	—	—	—	0.21	—	0.10	0.68	0.66	0.88	1.00														
10 10	—	—	—	0.15	—	—	0.52	0.57	0.91	0.88	1.00													
10 11	—	—	—	0.27	—	—	0.60	0.69	0.85	0.90	0.92	1.00												
10 12	—	—	—	0.20	—	—	0.51	0.67	0.79	0.88	0.86	0.92	1.00											
10 13	0.14	0.15	0.09	0.20	0.20	0.26	0.37	0.38	0.21	0.15	0.14	0.23	0.19	1.00										
10 14	—	—	—	0.13	—	0.13	0.31	0.42	0.48	0.39	0.52	0.48	0.36	0.34	1.00									
10 15	—	—	—	—	—	0.38	0.24	—	0.20	0.16	0.15	—	—	0.40	1.00									
10 16	—	—	—	—	0.17	—	0.10	0.48	0.59	0.62	0.59	0.68	0.54	0.42	0.68	0.36	1.00							
10 17	—	—	—	—	—	—	0.22	0.37	0.34	0.35	0.30	0.35	0.47	0.40	0.24	0.61	1.00							
10 18	0.08	—	—	—	—	—	0.13	0.30	0.44	0.36	0.30	0.23	0.27	0.08	0.16	0.14	0.29	0.39	1.00					
10 19	—	—	0.09	—	—	—	0.25	0.31	0.35	0.41	0.45	0.39	0.27	0.58	0.26	0.69	0.62	0.22	1.00					
10 20	—	—	0.32	—	—	—	0.35	0.39	0.37	0.43	0.53	0.55	0.41	0.20	0.62	—	0.70	0.32	—	0.74	1.00			
10 21	—	—	0.30	—	—	—	0.66	0.53	0.76	0.76	0.81	0.82	0.77	—	0.21	—	0.41	—	0.22	0.34	1.00			
10 22	—	—	0.13	—	—	0.16	0.35	0.26	0.15	0.23	0.16	—	—	—	—	—	—	—	—	—	—	1.00		
10 23	—	—	—	—	0.13	0.45	0.18	—	—	—	—	—	—	—	—	—	0.12	—	0.22	0.19	—	0.49	1.00	

Glacier Bay Oceanographic Monitoring Program Analysis of Observations, 1993–2009
Appendix G. Tables of temperature and salinity correlations between stations at select depth levels.

DEPTH STN	00	01	02	03	04	05	06	07	08	09	10	11	12	13	14	15	16	17	18	19	20	21	22	23
20 00	1.00																							
20 01	0.10	1.00																						
20 02	0.13	0.37	1.00																					
20 03	0.12	0.22	0.48	1.00																				
20 04	0.11	0.18	0.27	0.43	1.00																			
20 05	0.08	0.17	0.15	0.36	0.73	1.00																		
20 06	---	0.13	0.08	0.29	0.44	0.59	1.00																	
20 07	0.10	---	0.27	0.37	0.44	0.62	1.00																	
20 08	---	---	0.12	0.25	0.36	0.56	0.82	1.00																
20 09	---	---	0.13	0.21	0.27	0.53	0.72	0.90	1.00															
20 10	---	---	0.08	0.17	0.25	0.51	0.67	0.88	0.86	1.00														
20 11	---	---	0.10	0.09	0.16	0.47	0.55	0.70	0.57	0.78	1.00													
20 12	---	---	0.11	0.14	0.22	0.46	0.51	0.71	0.72	0.76	0.68	1.00												
20 13	---	0.14	0.12	0.21	0.49	0.46	0.51	0.62	0.55	0.35	0.41	0.34	0.32	1.00										
20 14	0.23	---	---	0.12	0.31	0.29	0.53	0.52	0.61	0.48	0.53	0.43	0.36	0.39	1.00									
20 15	0.16	---	---	---	0.27	---	0.26	0.18	0.16	0.19	---	---	0.15	0.49	1.00									
20 16	0.09	---	---	0.16	0.22	0.24	0.48	0.76	0.79	0.74	0.53	0.52	0.50	0.67	0.16	1.00								
20 17	0.08	0.09	---	0.11	0.25	0.35	0.44	0.59	0.64	0.47	0.52	0.46	0.32	0.64	0.56	0.41	0.61	1.00						
20 18	---	---	---	0.11	0.19	0.18	0.36	0.69	0.68	0.66	0.60	0.38	0.45	0.47	0.56	0.23	0.76	0.61	1.00					
20 19	---	---	---	---	0.12	0.13	0.25	0.59	0.74	0.77	0.66	0.48	0.52	0.42	0.57	0.28	0.76	0.58	0.67	1.00				
20 20	0.12	---	---	---	0.10	0.15	0.45	0.56	0.52	0.44	0.37	0.26	0.36	0.44	---	0.53	0.48	0.41	0.83	1.00				
20 21	---	---	---	---	---	0.21	0.33	0.36	0.47	0.57	0.35	0.55	---	---	0.54	---	---	---	---	---	1.00			
20 22	---	---	---	0.19	0.37	0.41	0.21	---	---	---	---	---	0.23	---	---	---	---	---	---	---	0.25	1.00		
20 23	---	---	0.15	0.35	0.45	0.49	0.30	---	---	---	---	0.44	---	---	---	---	---	---	---	---	---	0.64	1.00	

DEPTH STN	00	01	02	03	04	05	06	07	08	09	10	11	12	13	14	15	16	17	18	19	20	21	22	23
30 00	1.00																							
30 01	0.08	1.00																						
30 02	0.08	0.33	1.00																					
30 03	0.07	0.12	0.44	1.00																				
30 04	---	0.08	0.16	0.43	1.00																			
30 05	---	0.11	0.24	0.51	0.70	1.00																		
30 06	---	0.11	0.18	0.49	0.58	0.59	1.00																	
30 07	---	0.10	0.14	0.48	0.59	0.57	0.76	1.00																
30 08	---	0.12	0.15	0.40	0.55	0.54	0.73	0.81	1.00															
30 09	---	0.10	0.16	0.37	0.40	0.34	0.60	0.66	0.79	1.00														
30 10	---	0.09	0.18	0.31	0.43	0.34	0.60	0.69	0.82	0.92	1.00													
30 11	---	0.14	0.18	0.34	0.28	0.24	0.61	0.68	0.82	0.84	0.87	1.00												
30 12	---	0.13	0.17	0.34	0.28	0.24	0.52	0.56	0.67	0.76	0.76	0.83	1.00											
30 13	---	0.15	0.21	0.44	0.62	0.59	0.59	0.80	0.64	0.49	0.49	0.50	0.39	1.00										
30 14	---	0.08	0.07	0.23	0.40	0.28	0.56	0.54	0.56	0.60	0.57	0.51	0.43	0.39	1.00									
30 15	---	---	---	0.25	0.15	0.26	0.16	0.18	---	---	0.19	0.19	0.18	0.50	1.00									
30 16	---	0.12	0.19	0.42	0.48	0.42	0.62	0.76	0.73	0.81	0.70	0.63	0.60	0.65	0.17	1.00								
30 17	---	0.23	0.19	0.40	0.62	0.56	0.64	0.74	0.77	0.61	0.65	0.62	0.45	0.68	0.53	0.28	0.74	1.00						
30 18	---	0.09	0.12	0.39	0.53	0.37	0.59	0.77	0.75	0.72	0.74	0.69	0.56	0.57	0.61	0.25	0.85	0.79	1.00					
30 19	---	0.16	0.20	0.38	0.47	0.34	0.50	0.62	0.71	0.79	0.80	0.79	0.60	0.63	0.56	0.32	0.77	0.68	0.77	1.00				
30 20	---	0.14	0.14	0.41	0.27	0.32	0.37	0.52	0.50	0.59	0.55	0.56	0.34	0.54	0.38	0.21	0.58	0.48	0.51	0.83	1.00			
30 21	0.26	---	---	0.31	---	0.26	0.38	0.41	0.38	0.49	0.47	0.50	0.53	0.19	---	---	0.35	0.28	0.24	0.32	0.43	1.00		
30 22	---	---	0.19	0.49	0.50	0.71	0.56	0.36	0.22	0.21	0.17	0.18	0.44	0.21	---	0.21	0.34	0.13	---	0.30	1.00			
30 23	---	---	0.25	0.35	0.53	0.57	0.70	0.61	0.44	0.29	0.32	0.20	0.24	0.53	0.27	---	0.38	0.45	0.23	0.20	0.28	0.38	0.73	1.00

Glacier Bay Oceanographic Monitoring Program Analysis of Observations, 1993–2009
Appendix G. Tables of temperature and salinity correlations between stations at select depth levels.

DEPTH STN	00	01	02	03	04	05	06	07	08	09	10	11	12	13	14	15	16	17	18	19	20	21	22	23
50 00	1.00																							
50 01	0.44	1.00																						
50 02	---	---	1.00																					
50 03	---	---	0.50	1.00																				
50 04	---	---	0.24	0.57	1.00																			
50 05	---	---	0.29	0.58	0.80	1.00																		
50 06	---	---	0.23	0.58	0.78	0.82	1.00																	
50 07	---	---	0.24	0.57	0.77	0.80	0.89	1.00																
50 08	---	---	0.22	0.55	0.75	0.76	0.87	0.89	1.00															
50 09	---	---	0.23	0.49	0.58	0.59	0.75	0.80	0.88	1.00														
50 10	---	---	0.21	0.48	0.61	0.63	0.74	0.80	0.90	0.90	1.00													
50 11	---	---	0.26	0.54	0.59	0.66	0.77	0.78	0.89	0.89	0.94	1.00												
50 12	---	---	0.13	0.34	0.32	0.43	0.45	0.49	0.60	0.64	0.70	0.75	1.00											
50 13	---	---	0.33	0.65	0.80	0.77	0.86	0.88	0.81	0.70	0.67	0.66	0.35	1.00										
50 14	---	---	0.15	0.43	0.47	0.54	0.57	0.62	0.62	0.65	0.66	0.63	0.37	0.63	1.00									
50 15	---	---	---	0.28	0.42	0.47	0.49	0.55	0.45	0.34	0.32	0.27	---	0.48	0.37	1.00								
50 16	---	---	0.24	0.58	0.73	0.73	0.81	0.84	0.85	0.78	0.82	0.81	0.49	0.83	0.73	0.50	1.00							
50 17	---	---	0.23	0.52	0.78	0.75	0.84	0.83	0.85	0.76	0.77	0.77	0.47	0.85	0.66	0.50	0.89	1.00						
50 18	---	---	0.14	0.56	0.76	0.72	0.83	0.84	0.83	0.79	0.78	0.77	0.48	0.85	0.71	0.49	0.92	0.92	1.00					
50 19	---	---	0.34	0.64	0.76	0.80	0.81	0.80	0.83	0.72	0.78	0.78	0.50	0.81	0.72	0.51	0.86	0.88	0.87	1.00				
50 20	---	---	0.23	0.40	0.42	0.48	0.48	0.44	0.41	0.43	0.42	0.18	0.43	0.34	0.69	0.50	0.43	0.43	0.82	1.00				
50 21	---	---	0.26	0.31	0.35	0.43	0.50	0.44	0.51	0.54	0.51	0.54	0.63	0.36	0.20	---	0.44	0.43	0.37	0.47	0.25	1.00		
50 22	---	---	0.24	0.43	0.64	0.72	0.75	0.69	0.68	0.52	0.55	0.65	0.45	0.63	0.47	0.62	0.58	0.69	0.62	0.72	0.31	0.50	1.00	
50 23	---	---	0.35	0.44	0.59	0.68	0.73	0.70	0.66	0.48	0.53	0.68	0.44	0.60	0.87	0.60	0.70	0.57	0.76	0.24	0.50	0.93	1.00

DEPTH STN	00	01	02	03	04	05	06	07	08	09	10	11	12	13	14	15	16	17	18	19	20	21	22	23
75 00	...																							
75 01	...	1.00																						
75 02	...	0.50	1.00																					
75 03	...	0.29	0.55	1.00																				
75 04	...	0.40	0.63	0.83	1.00																			
75 05	...	0.37	0.55	0.83	0.88	1.00																		
75 06	...	0.44	0.59	0.81	0.84	0.91	1.00																	
75 07	...	0.35	0.56	0.82	0.84	0.89	0.91	1.00																
75 08	...	0.37	0.49	0.77	0.75	0.82	0.86	0.91	1.00															
75 09	...	0.36	0.53	0.75	0.76	0.85	0.86	0.83	0.90	0.94	1.00													
75 10	...	0.41	0.51	0.73	0.71	0.78	0.78	0.86	0.83	0.88	1.00													
75 11	...	0.45	0.47	0.61	0.68	0.69	0.67	0.75	0.74	0.81	0.78	1.00												
75 12	...	0.42	0.70	0.86	0.85	0.88	0.89	0.87	0.81	0.78	0.76	0.61	1.00											
75 13													1.00											
75 14	...	---	0.15	0.37	0.37	0.43	0.53	0.34	0.30	0.26	0.20	---	0.34	---	1.00									
75 15	...	0.34	0.54	0.77	0.73	0.80	0.83	0.82	0.83	0.79	0.64	0.81	---	---	0.35	1.00								
75 16	...	0.17	0.40	0.76	0.77	0.81	0.78	0.81	0.77	0.75	0.57	0.70	0.77	---	0.42	0.81	1.00							
75 17	...	0.30	0.55	0.75	0.78	0.85	0.86	0.85	0.84	0.71	0.84	---	0.34	0.90	0.85	1.00								
75 18	...	0.42	0.55	0.74	0.85	0.82	0.76	0.81	0.78	0.81	0.80	0.72	0.80	0.48	0.36	0.78	0.82	0.86	1.00					
75 19	...	0.37	0.35	0.41	0.56	0.50	0.47	0.49	0.45	0.57	0.40	0.38	0.49	---	0.56	0.46	0.45	0.53	0.80	1.00				
75 20	...	0.37	0.34	0.37	0.49	0.50	0.43	0.45	0.57	0.53	0.37	0.73	0.34	...	0.38	0.31	0.38	0.54	0.38	1.00				
75 21	...	0.36	0.41	0.65	0.80	0.73	0.71	0.74	0.69	0.70	0.71	0.77	0.70	0.93	---	0.51	0.65	0.69	0.77	0.44	0.58	1.00		
75 22	...		0.00	0.00	0.00	0.00	0.00	0.00	0.00	0.00	0.00	0.00	0.00	0.00	0.00	0.00	0.00	0.00	0.00	0.00	0.00	0.00	...	
75 23	...																						0.00	...

165

Glacier Bay Oceanographic Monitoring Program Analysis of Observations, 1993–2009
Appendix G. Tables of temperature and salinity correlations between stations at select depth levels.

DEPTH STN	00	01	02	03	04	05	06	07	08	09	10	11	12	13	14	15	16	17	18	19	20	21	22	23
100 00	...																							
100 01	...																							
100 02	...																							
100 03	...	1.00																						
100 04	...	0.63	1.00																					
100 05	...	0.73	0.83	1.00																				
100 06	...	0.71	0.83	0.85	1.00																			
100 07	...	0.72	0.81	0.85	0.91	1.00																		
100 08	...	0.67	0.80	0.86	0.92	0.93	1.00																	
100 09	...	0.66	0.80	0.78	0.87	0.89	0.93	1.00																
100 10	...	0.59	0.78	0.76	0.86	0.89	0.91	0.96	1.00															
100 11	...	0.64	0.75	0.81	0.82	0.82	0.87	0.87	0.92	1.00														
100 12	...	0.38	0.52	0.61	0.63	0.64	0.68	0.71	0.73	0.82	1.00													
100 13	...	0.75	0.81	0.81	0.85	0.85	0.86	0.77	0.75	0.79	0.52	1.00												
100 14	...																							
100 15	...										0.42	0.37	---	---	1.00									
100 16	...	0.52	0.60	0.64	0.73	0.72	0.75	0.73	0.78	0.77	0.53	0.70	...	...	0.36	1.00								
100 17	...	0.46	0.62	0.70	0.76	0.75	0.79	0.78	0.77	0.78	0.55	0.70	...	...	0.43	0.80	1.00							
100 18	...	0.36	0.52	0.63	0.69	0.68	0.71	0.68	0.71	0.80	0.67	0.70	...	...	0.50	0.81	0.82	1.00						
100 19	...	0.52	0.59	0.74	0.70	0.65	0.72	0.67	0.70	0.79	0.62	0.69	...	...	0.57	0.78	0.79	0.81	1.00					
100 20	...	---	0.30	0.39	0.35	0.36	0.41	0.35	0.35	0.46	0.39	0.44	...	...	0.41	0.56	0.43	0.53	0.88	1.00				
100 21	...	---	0.53	0.47	0.52	0.49	0.53	0.61	0.49	0.54	0.77	0.35	...	...	---	0.38	0.36	0.35	0.36	0.25	1.00			
100 22	...																							
100 23	...	0.74	0.68	0.76	0.72	0.70	0.75	0.70	0.71	0.77	0.68	0.66	...	...	0.48	0.59	0.59	0.67	0.26	0.58	1.00			

DEPTH STN	00	01	02	03	04	05	06	07	08	09	10	11	12	13	14	15	16	17	18	19	20	21	22	23
150 00	...																							
150 01	...																							
150 02	...																							
150 03	...	1.00																						
150 04	...	0.81	1.00																					
150 05	...	0.85	0.86	1.00																				
150 06	...	0.86	0.79	0.92	1.00																			
150 07	...	0.82	0.84	0.92	0.92	1.00																		
150 08	...	0.75	0.80	0.85	0.88	0.93	1.00																	
150 09	...	0.76	0.80	0.89	0.90	0.92	0.93	1.00																
150 10	...	0.70	0.83	0.83	0.81	0.86	0.88	0.93	1.00															
150 11	...	0.40	0.59	0.56	0.56	0.60	0.68	0.75	0.75	1.00														
150 12	...																							
150 13	...																							
150 14	...																							
150 15	...														1.00									
150 16	...	0.61	0.62	0.69	0.66	0.67	0.69	0.73	0.72	0.54	...	...	...	0.85	1.00									
150 17	...	0.56	0.62	0.73	0.63	0.68	0.70	0.71	0.74	0.56	...	...	...	0.91	0.87	1.00								
150 18	...	0.53	0.59	0.66	0.62	0.67	0.71	0.74	0.76	0.66	...	...	...	0.86	0.82	0.94	1.00							
150 19	...	0.50	0.60	0.61	0.61	0.67	0.74	0.72	0.74	0.75	...	...	...	0.47	0.31	0.48	0.72	1.00						
150 20	...	0.21	0.19	0.21	0.26	0.24	0.25	0.26	0.29	0.25	...	...	...	0.45	0.35	0.51	0.59	0.28	1.00					
150 21	...	0.36	0.51	0.48	0.53	0.53	0.64	0.58	0.61	0.80	...	...	...	---	---	---	---	0.87	1.00					
150 22	...	---	0.88	---	0.80	0.96	0.74	0.93	0.94	...	...	...	...	...	...	...	...	...	...					
150 23	...																							

Glacier Bay Oceanographic Monitoring Program Analysis of Observations, 1993–2009
Appendix G. Tables of temperature and salinity correlations between stations at select depth levels.

DEPTH STN	00	01	02	03	04	05	06	07	08	09	10	11	12	13	14	15	16	17	18	19	20	21	22	23
200 00	...																							
200 01	...	...																						
200 02	...	...	...																					
200 03	...	...	...	...																				
200 04	...	...	...	...	1.00																			
200 05	...	...	...	...	0.84	1.00																		
200 06	...	...	...	...	0.87	0.83	1.00																	
200 07	...	...	...	...	0.84	0.82	0.92	1.00																
200 08	...	...	...	...	0.83	0.81	0.93	0.93	1.00															
200 09	...	...	...	...	0.84	0.85	0.90	0.91	0.94	1.00														
200 10	...	...	...	...	0.82	0.78	0.88	0.89	0.90	0.92	1.00													
200 11	...	...	...	...	0.80	0.81	0.86	0.87	0.90	0.92	0.92	1.00												
200 12	...	...	...	...	0.51	0.59	0.60	0.68	0.59	0.69	0.76	0.75	1.00											
200 13	...	...	...	...	...	...	...	...	...	...	...	...	...											
200 14	...	...	...	...	...	...	...	...	...	...	...	...	...	...										
200 15	...	...	...	...	0.75	0.73	0.81	0.78	0.77	0.81	0.83	0.83	0.63	...	...	1.00								
200 16	...	...	...	...	0.00	0.00	0.00	0.00	0.00	0.00	-0.00	0.00	0.00	...	...	...	0.00	...						
200 17	...	...	...	...	...	...	...	...	...	...	...	...	...	...	...	...	...	1.00						
200 18	...	...	...	...	0.66	0.59	0.68	0.65	0.66	0.80	0.76	0.77	0.82	...	...	...	...	0.82	...	1.00				
200 19	...	...	...	...	...	...	...	...	...	...	...	...	...	...	...	...	...	...	...	...				
200 20	...	...	...	...	...	...	...	...	...	...	...	...	...	...	...	...	...	...	...	...	1.00			
200 21	...	...	...	...	...	...	...	...	...	...	...	...	...	...	...	...	...	...	...	...	...	...		
200 22	...	...	...	...	...	...	...	...	...	...	...	...	...	...	...	...	...	...	...	...	...	...	...	
200 23	...	...	...	...	...	...	...	...	...	...	...	...	...	...	...	...	...	...	...	...	...	...	...	...

DEPTH STN	00	01	02	03	04	05	06	07	08	09	10	11	12	13	14	15	16	17	18	19	20	21	22	23
250 00	...																							
250 01	...	...																						
250 02	...	...	...																					
250 03	...	...	...	...																				
250 04	...	...	...	...	1.00																			
250 05	...	...	...	...	0.85	1.00																		
250 06	...	...	...	...	0.82	0.86	1.00																	
250 07	...	...	...	...	0.88	0.86	0.91	1.00																
250 08	...	...	...	...	0.87	0.88	0.91	0.94	1.00															
250 09	...	...	...	...	0.87	0.88	0.92	0.91	0.97	1.00														
250 10	...	...	...	...	0.80	0.81	0.87	0.90	0.94	0.96	1.00													
250 11	...	...	...	...	0.84	0.84	0.84	0.87	0.93	0.95	0.97	1.00												
250 12	...	...	...	...	0.57	0.54	0.57	0.54	0.68	0.72	0.78	0.70	1.00											
250 13	...	...	...	...	...	...	...	...	...	...	...	...	...											
250 14	...	...	...	...	...	...	...	...	...	...	...	...	...	...										
250 15	...	...	...	...	...	...	...	...	...	...	...	...	...	...	...									
250 16	...	...	...	...	0.77	0.75	0.77	0.78	0.84	0.86	0.87	0.84	0.57	...	...	...	1.00							
250 17	...	...	...	...	...	...	...	...	...	...	...	...	...	...	...	...	...	...						
250 18	...	...	...	...	...	...	...	...	...	...	...	...	...	...	...	...	...	...	...					
250 19	...	...	...	...	...	...	...	...	...	...	...	...	...	...	...	...	...	...	...	...				
250 20	...	...	...	...	...	...	...	...	...	...	...	...	...	...	...	...	...	...	...	...	...			
250 21	...	...	...	...	...	...	...	...	...	...	...	...	...	...	...	...	...	...	...	...	...	...		
250 22	...	...	...	...	...	...	...	...	...	...	...	...	...	...	...	...	...	...	...	...	...	...	...	
250 23	...	...	...	...	...	...	...	...	...	...	...	...	...	...	...	...	...	...	...	...	...	...	...	...

Glacier Bay Oceanographic Monitoring Program Analysis of Observations, 1993–2009
Appendix G. Tables of temperature and salinity correlations between stations at select depth levels.

Correlation of temperature at constant depth levels between station pairs; "—" denotes no significant relation found; "..." denotes no data taken at that depth level.

DEPTH STN	00	01	02	03	04	05	06	07	08	09	10	11	12	13	14	15	16	17	18	19	20	21	22	23
00 00	1.00																							
00 01	0.32	1.00																						
00 02	0.29	0.55	1.00																					
00 03	0.17	0.49	0.54	1.00																				
00 04	—	0.13	0.13	0.20	1.00																			
00 05	—	0.14	0.08	0.15	0.53	1.00																		
00 06	—	0.16	0.17	0.35	0.40	0.59	1.00																	
00 07	—	0.08	—	0.23	0.29	0.43	0.63	1.00																
00 08	—	0.13	0.11	0.15	0.33	0.48	0.50	0.56	1.00															
00 09	—	0.11	0.13	0.30	0.17	0.27	0.47	0.39	0.59	1.00														
00 10	—	0.11	0.10	0.16	0.22	0.34	0.47	0.44	0.59	0.65	1.00													
00 11	—	—	—	—	0.09	0.16	0.18	0.27	0.40	0.49	1.00													
00 12	—	—	0.15	—	—	—	—	—	—	0.18	1.00													
00 13	—	0.14	0.10	0.23	0.37	0.43	0.58	0.55	0.51	0.36	0.39	0.14	—	1.00										
00 14	—	0.08	—	0.10	0.25	0.48	0.63	0.59	0.50	0.37	0.45	0.21	—	0.73	1.00									
00 15	—	0.13	—	0.14	—	0.30	0.18	—	—	0.26	0.32	—	—	0.48	0.35	1.00								
00 16	0.14	—	—	0.07	0.30	0.34	0.48	0.37	0.36	0.15	0.27	—	—	0.61	0.66	0.38	1.00							
00 17	—	—	—	—	0.09	0.19	0.35	0.42	0.35	0.31	0.13	0.25	—	0.55	0.55	0.31	0.66	1.00						
00 18	—	—	—	0.08	0.14	0.16	0.18	0.23	0.18	—	0.10	—	—	0.33	0.25	—	0.39	0.56	1.00					
00 19	—	—	—	—	0.09	—	—	—	—	—	—	—	—	0.11	0.33	0.35	1.00							
00 20	—	—	0.12	—	0.13	0.10	0.09	—	0.14	—	—	—	—	—	0.13	0.12	0.31	1.00						
00 21	—	—	—	—	—	—	—	0.17	—	—	0.42	—	—	—	—	—	0.26	1.00						
00 22	—	0.24	0.21	0.28	0.20	0.23	0.32	0.42	0.29	0.19	—	—	—	0.14	0.30	—	—	0.66	1.00					
00 23	—	—	0.15	—	0.18	0.25	0.24	0.35	0.28	—	—	—	—	0.28	—	—	—	—	—	—	—	—	—	1.00

DEPTH STN	00	01	02	03	04	05	06	07	08	09	10	11	12	13	14	15	16	17	18	19	20	21	22	23
10 00	1.00																							
10 01	0.48	1.00																						
10 02	0.44	0.70	1.00																					
10 03	0.39	0.61	0.78	1.00																				
10 04	0.29	0.35	0.39	0.58	1.00																			
10 05	0.24	0.35	0.47	0.49	0.58	1.00																		
10 06	0.27	0.51	0.53	0.57	0.50	0.61	1.00																	
10 07	0.13	0.21	0.23	0.22	0.39	0.40	0.72	1.00																
10 08	0.27	0.27	0.31	0.36	0.42	0.40	0.61	0.60	1.00															
10 09	0.36	0.36	0.42	0.49	0.39	0.40	0.65	0.14	0.57	1.00														
10 10	0.17	0.19	0.24	0.24	0.20	0.23	0.30	0.40	0.67	0.68	1.00													
10 11	0.14	0.18	0.23	0.19	0.09	0.09	0.23	—	0.15	0.53	0.47	0.64	1.00											
10 12	0.10	0.12	0.15	0.12	—	0.11	0.26	0.16	0.32	0.43	0.49	0.72	1.00											
10 13	0.36	0.40	0.54	0.49	0.70	0.76	0.56	0.49	0.48	0.37	0.26	0.11	0.08	1.00										
10 14	0.37	0.43	0.53	0.56	0.68	0.59	0.54	0.32	0.44	0.35	0.19	—	—	0.76	1.00									
10 15	0.27	0.17	0.21	0.35	0.42	0.49	0.46	0.26	0.16	—	—	0.20	—	0.60	0.65	1.00								
10 16	0.28	0.31	0.42	0.46	0.62	0.67	0.55	0.41	0.46	0.27	—	0.12	0.81	0.79	0.68	1.00								
10 17	0.30	0.31	0.40	0.41	0.54	0.68	0.61	0.38	0.41	0.42	0.29	0.10	0.21	0.75	0.65	0.56	0.75	1.00						
10 18	0.30	0.34	0.41	0.41	0.36	0.45	0.52	0.23	0.32	0.45	0.29	0.16	0.18	0.49	0.48	0.21	0.56	0.71	1.00					
10 19	0.18	0.18	0.28	0.29	0.25	0.39	0.36	0.12	0.25	0.46	0.25	0.24	0.32	0.45	0.42	0.35	0.15	0.45	0.63	0.65	1.00			
10 20	0.22	0.30	0.42	0.41	0.26	0.39	0.55	0.24	0.33	0.45	0.29	0.32	0.41	0.51	0.49	0.33	0.59	0.54	0.76	1.00				
10 21	—	0.18	0.24	0.25	—	0.27	0.30	0.47	0.62	0.63	0.71	0.80	0.18	—	0.21	0.19	0.39	0.41	1.00					
10 22	0.26	0.51	0.51	0.45	0.42	0.64	0.80	0.66	0.60	0.63	0.31	0.26	0.29	0.62	0.60	—	0.71	0.68	0.63	0.40	0.43	0.31	1.00	
10 23	0.31	0.36	0.40	0.31	0.43	0.61	0.73	0.62	0.67	0.68	0.38	0.34	0.45	0.56	0.47	—	0.59	0.67	0.57	0.50	0.36	0.45	0.83	1.00

Glacier Bay Oceanographic Monitoring Program Analysis of Observations, 1993–2009
Appendix G. Tables of temperature and salinity correlations between stations at select depth levels.

DEPTH	STN	00	01	02	03	04	05	06	07	08	09	10	11	12	13	14	15	16	17	18	19	20	21	22	23
20	00	1.00																							
20	01	0.52	1.00																						
20	02	0.50	0.77	1.00																					
20	03	0.46	0.78	0.80	1.00																				
20	04	0.33	0.57	0.59	0.77	1.00																			
20	05	0.32	0.46	0.46	0.69	0.72	1.00																		
20	06	0.30	0.50	0.47	0.61	0.62	0.75	1.00																	
20	07	0.37	0.55	0.54	0.58	0.64	0.65	0.80	1.00																
20	08	0.31	0.39	0.42	0.51	0.64	0.61	0.76	0.83	1.00															
20	09	0.33	0.42	0.49	0.49	0.58	0.67	0.80	0.84	0.88	1.00														
20	10	0.30	0.37	0.41	0.44	0.49	0.61	0.73	0.80	0.81	0.89	1.00													
20	11	0.20	0.20	0.27	0.25	0.33	0.31	0.43	0.49	0.57	0.59	0.68	1.00												
20	12	0.18	0.12	0.13	0.16	0.15	0.28	0.42	0.42	0.40	0.43	0.53	0.68	1.00											
20	13	0.31	0.46	0.51	0.63	0.77	0.69	0.63	0.65	0.59	0.57	0.46	0.23	0.16	1.00										
20	14	0.46	0.57	0.65	0.66	0.74	0.64	0.64	0.68	0.66	0.65	0.51	0.25	0.15	0.82	1.00									
20	15	0.29	0.14	0.15	0.29	0.45	0.50	0.34	0.38	0.46	0.45	0.26	---	0.59	0.80	1.00									
20	16	0.32	0.30	0.37	0.48	0.67	0.72	0.68	0.64	0.71	0.69	0.63	0.30	0.14	0.71	0.70	0.50	1.00							
20	17	0.28	0.22	0.26	0.37	0.49	0.66	0.66	0.60	0.65	0.69	0.64	0.27	0.18	0.57	0.55	0.53	0.84	1.00						
20	18	0.31	0.28	0.31	0.42	0.49	0.70	0.66	0.64	0.64	0.74	0.73	0.46	0.30	0.50	0.46	0.33	0.73	0.85	1.00					
20	19	0.32	0.17	0.25	0.28	0.33	0.49	0.50	0.46	0.54	0.61	0.63	0.42	0.36	0.32	0.33	0.17	0.45	0.64	0.79	1.00				
20	20	0.27	0.29	0.32	0.38	0.42	0.47	0.64	0.59	0.58	0.66	0.71	0.54	0.41	0.48	0.41	0.21	0.54	0.64	0.80	0.84	1.00			
20	21	---	---	0.18	0.20	0.25	0.22	0.25	0.32	0.41	0.39	0.35	0.49	0.72	0.32	0.29	---	0.22	0.24	0.26	0.45	0.36	1.00		
20	22	0.42	0.56	0.54	0.63	0.69	0.63	0.79	0.74	0.71	0.72	0.68	0.58	0.48	0.69	0.68	---	0.76	0.64	0.70	0.39	0.49	0.51	1.00	
20	23	0.30	0.41	0.44	0.47	0.44	0.55	0.70	0.65	0.74	0.73	0.68	0.56	0.62	0.60	0.63	---	0.66	0.68	0.60	0.67	0.58	0.77	1.00	

DEPTH	STN	00	01	02	03	04	05	06	07	08	09	10	11	12	13	14	15	16	17	18	19	20	21	22	23
30	00	1.00																							
30	01	0.58	1.00																						
30	02	0.45	0.87	1.00																					
30	03	0.42	0.81	0.83	1.00																				
30	04	0.38	0.67	0.69	0.80	1.00																			
30	05	0.33	0.50	0.47	0.66	0.72	1.00																		
30	06	0.29	0.52	0.53	0.56	0.69	0.72	1.00																	
30	07	0.31	0.50	0.54	0.53	0.63	0.62	0.82	1.00																
30	08	0.22	0.44	0.47	0.50	0.63	0.61	0.75	0.74	1.00															
30	09	0.28	0.45	0.49	0.51	0.63	0.70	0.81	0.78	0.85	1.00														
30	10	0.25	0.40	0.41	0.40	0.59	0.60	0.72	0.79	0.88	0.86	1.00													
30	11	0.29	0.31	0.31	0.46	0.31	0.50	0.58	0.79	0.67	0.67	0.77	1.00												
30	12	0.10	0.18	0.13	0.15	0.21	0.16	0.34	0.44	0.50	0.53	0.54	0.71	1.00											
30	13	0.31	0.53	0.54	0.55	0.71	0.60	0.65	0.69	0.53	0.65	0.50	0.26	0.21	1.00										
30	14	0.41	0.66	0.72	0.65	0.67	0.62	0.68	0.68	0.57	0.70	0.55	0.30	0.20	0.81	1.00									
30	15	0.33	0.24	0.25	0.38	0.36	0.44	0.43	0.41	0.29	0.52	0.28	---	0.53	0.76	1.00									
30	16	0.29	0.43	0.46	0.52	0.66	0.69	0.71	0.67	0.65	0.74	0.71	0.51	0.28	0.63	0.63	0.42	1.00							
30	17	0.25	0.30	0.31	0.37	0.56	0.68	0.75	0.63	0.75	0.76	0.79	0.50	0.28	0.48	0.49	0.37	0.82	1.00						
30	18	0.19	0.32	0.39	0.53	0.65	0.69	0.68	0.68	0.81	0.73	0.81	0.60	0.40	0.44	0.47	0.24	0.70	0.89	1.00					
30	19	0.20	0.18	0.19	0.24	0.39	0.47	0.53	0.48	0.61	0.60	0.62	0.47	0.47	0.30	0.30	0.17	0.46	0.63	0.76	1.00				
30	20	0.20	0.31	0.30	0.34	0.46	0.45	0.64	0.59	0.64	0.66	0.71	0.64	0.58	0.46	0.40	0.23	0.58	0.62	0.75	0.85	1.00			
30	21	---	---	0.17	0.21	0.33	0.29	0.26	0.31	0.44	0.45	0.42	0.63	0.74	0.45	0.29	---	0.28	0.28	0.36	0.55	0.48	1.00		
30	22	0.37	0.60	0.53	0.63	0.69	0.63	0.53	0.60	0.67	0.54	0.53	0.47	0.58	0.65	---	0.72	0.59	0.55	0.30	0.34	0.51	1.00		
30	23	0.24	0.49	0.50	0.54	0.50	0.64	0.67	0.65	0.74	0.76	0.71	0.68	0.61	0.60	0.68	---	0.70	0.70	0.74	0.62	0.69	0.64	0.75	1.00

Glacier Bay Oceanographic Monitoring Program Analysis of Observations, 1993–2009
Appendix G. Tables of temperature and salinity correlations between stations at select depth levels.

DEPTH STN	00	01	02	03	04	05	06	07	08	09	10	11	12	13	14	15	16	17	18	19	20	21	22	23
50 00	1.00																							
50 01	0.70	1.00																						
50 02	0.44	0.64	1.00																					
50 03	0.37	0.54	0.84	1.00																				
50 04	0.32	0.45	0.69	0.76	1.00																			
50 05	0.26	0.35	0.51	0.62	0.71	1.00																		
50 06	0.41	0.27	0.56	0.56	0.72	0.81	1.00																	
50 07	0.50	0.25	0.52	0.51	0.70	0.72	0.86	1.00																
50 08	0.53	0.19	0.47	0.50	0.67	0.69	0.82	0.81	1.00															
50 09	0.52	0.17	0.42	0.45	0.59	0.68	0.77	0.76	0.88	1.00														
50 10	0.54	—	0.34	0.38	0.54	0.60	0.74	0.75	0.87	0.90	1.00													
50 11	—	—	0.28	0.31	0.44	0.40	0.58	0.59	0.75	0.74	0.84	1.00												
50 12	—	—	0.11	0.14	0.31	0.22	0.37	0.45	0.56	0.58	0.61	0.74	1.00											
50 13	0.54	—	0.49	0.67	0.69	0.76	0.71	0.70	0.62	0.55	0.50	0.43	0.29	1.00										
50 14	0.61	—	0.56	0.67	0.68	0.58	0.59	0.62	0.62	0.54	0.51	0.47	0.36	0.29	0.80	1.00								
50 15	0.61	—	0.55	0.48	0.54	0.43	0.34	0.42	0.30	0.26	0.22	—	—	0.49	0.64	1.00								
50 16	0.43	—	0.37	0.55	0.55	0.67	0.70	0.73	0.79	0.72	0.74	0.60	0.37	0.61	0.56	0.32	1.00							
50 17	0.48	—	0.17	0.43	0.48	0.64	0.72	0.80	0.73	0.88	0.83	0.86	0.72	0.51	0.53	0.52	0.29	0.80	1.00					
50 18	0.36	—	0.15	0.42	0.45	0.59	0.68	0.75	0.74	0.86	0.80	0.74	0.52	0.53	0.49	0.20	0.78	0.94	1.00					
50 19	0.56	—	0.15	0.26	0.27	0.41	0.43	0.63	0.51	0.62	0.54	0.71	0.63	0.54	0.38	0.42	0.38	0.47	0.73	1.00				
50 20	0.63	—	0.22	0.24	0.29	0.47	0.43	0.61	0.57	0.66	0.62	0.77	0.72	0.66	0.48	0.43	0.36	0.60	0.74	0.76	0.87	1.00		
50 21	—	—	0.20	0.27	0.44	0.28	0.39	0.47	0.53	0.50	0.54	0.63	0.79	0.49	0.51	—	0.42	0.50	0.54	0.58	0.52	1.00		
50 22	—	—	0.27	0.55	0.59	0.63	0.75	0.69	0.73	0.78	0.69	0.56	0.45	0.69	0.75	—	0.78	0.79	0.78	0.47	0.52	0.50	1.00	
50 23	—	—	0.28	0.60	0.70	0.76	0.72	0.77	0.81	0.77	0.81	0.68	0.77	0.82	—	0.75	0.81	0.82	0.57	0.58	0.66	0.92	1.00

DEPTH STN	00	01	02	03	04	05	06	07	08	09	10	11	12	13	14	15	16	17	18	19	20	21	22	23
75 00	...																							
75 01	...	1.00																						
75 02	...	0.73	1.00																					
75 03	...	0.47	0.62	1.00																				
75 04	...	0.40	0.47	0.73	1.00																			
75 05	...	0.43	0.44	0.70	0.86	1.00																		
75 06	...	0.47	0.42	0.70	0.75	0.82	1.00																	
75 07	...	0.37	0.45	0.61	0.76	0.83	0.73	1.00																
75 08	...	0.39	0.39	0.55	0.72	0.82	0.75	0.90	1.00															
75 09	...	0.27	0.33	0.49	0.58	0.66	0.68	0.85	0.89	1.00														
75 10	...	0.14	0.29	0.38	0.43	0.56	0.55	0.78	0.75	0.83	1.00													
75 11	...	—	0.22	0.31	0.21	0.35	0.39	0.53	0.53	0.62	0.77	1.00												
75 12	...	0.59	0.72	0.81	0.78	0.76	0.78	0.63	0.56	0.52	0.43	0.30	1.00											
75 13	...	—	0.83	—	—	—	—	—	—	—	—	—	1.00											
75 14	...	0.31	0.43	0.42	0.16	0.23	0.24	0.18	—	0.23	—	—	0.63	—	1.00									
75 15	...	0.50	0.60	0.70	0.74	0.70	0.73	0.75	0.76	0.66	0.48	0.78	—	0.38	1.00									
75 16	...	0.29	0.47	0.58	0.70	0.68	0.64	0.79	0.72	0.74	0.66	0.45	0.66	—	0.55	0.76	1.00							
75 17	...	0.43	0.51	0.57	0.68	0.71	0.69	0.82	0.78	0.71	0.55	0.67	—	0.48	0.81	0.86	1.00							
75 18	...	0.23	0.31	0.42	0.45	0.51	0.46	0.57	0.57	0.69	0.62	0.55	0.49	—	0.54	0.61	0.74	0.78	1.00					
75 19	...	0.18	0.28	0.51	0.47	0.53	0.60	0.59	0.66	0.63	0.55	—	0.45	0.68	0.73	0.77	0.83	1.00						
75 20	...	—	0.33	0.39	0.36	0.51	0.52	0.58	0.55	0.70	0.79	0.36	...	—	0.53	0.48	0.56	0.51	0.47	1.00				
75 21	...	0.56	0.64	0.71	0.80	0.80	0.79	0.83	0.77	0.69	0.55	0.72	...	—	0.83	0.74	0.82	0.51	0.52	0.58	1.00			
75 22	...	0.00	0.00	0.00	0.00	0.00	0.00	0.00	0.00	0.00	0.00	0.00	-0.00	...	0.00	0.00	0.00	0.00	0.00	0.00	0.00	...	0.00	
75 23	...																							

Glacier Bay Oceanographic Monitoring Program Analysis of Observations, 1993–2009
Appendix G. Tables of temperature and salinity correlations between stations at select depth levels.

DEPTH STN	00	01	02	03	04	05	06	07	08	09	10	11	12	13	14	15	16	17	18	19	20	21	22	23
100 00	...																							
100 01	...	...																						
100 02	...	...	1.00																					
100 03	...	...	---	1.00																				
100 04	...	...	---	1.00																				
100 05	...	...	---	0.72	1.00																			
100 06	...	...	0.22	0.64	0.78	1.00																		
100 07	...	...	---	0.66	0.79	0.82	1.00																	
100 08	...	...	---	0.62	0.78	0.85	0.81	1.00																
100 09	...	...	---	0.57	0.71	0.79	0.79	0.92	1.00															
100 10	...	...	---	0.53	0.64	0.74	0.75	0.87	0.93	1.00														
100 11	...	...	---	0.40	0.40	0.54	0.57	0.71	0.77	0.86	1.00													
100 12	...	...	---	0.32	0.22	0.32	0.41	0.48	0.52	0.61	0.71	1.00												
100 13	...	...	0.51	0.77	0.71	0.69	0.71	0.58	0.50	0.52	0.32	0.29	1.00											
100 14	...	...																						
100 15	...	...	0.34	---	---	---	---	---	---	---	0.71	---	1.00											
100 16	...	...	---	0.74	0.75	0.68	0.73	0.73	0.70	0.69	0.54	0.45	0.76	...	0.27	1.00								
100 17	...	...	0.20	0.66	0.67	0.73	0.69	0.75	0.73	0.74	0.59	0.46	0.68	...	0.54	0.77	1.00							
100 18	...	...	---	0.69	0.68	0.70	0.73	0.72	0.69	0.71	0.52	0.41	0.74	...	0.39	0.81	0.87	1.00						
100 19	...	...	---	0.50	0.47	0.52	0.49	0.54	0.51	0.55	0.48	0.42	0.60	...	0.38	0.64	0.75	0.77	1.00					
100 20	...	...	---	0.60	0.49	0.55	0.56	0.55	0.47	0.53	0.41	0.46	0.66	...	0.41	0.76	0.71	0.67	0.84	1.00				
100 21	...	...	0.48	0.44	0.38	0.53	0.52	0.61	0.62	0.62	0.61	0.80	0.33	...	---	0.55	0.51	0.49	0.39	0.38	1.00			
100 22	...	...	0.37	0.62	0.72	0.76	0.79	0.85	0.75	0.80	0.66	0.49	0.63	...	---	0.74	0.77	0.77	0.51	0.52	0.57	1.00		
100 23																								

DEPTH STN	00	01	02	03	04	05	06	07	08	09	10	11	12	13	14	15	16	17	18	19	20	21	22	23
150 00																								
150 01	...																							
150 02	...	...																						
150 03	...	...	1.00																					
150 04	...	...	0.71	1.00																				
150 05	...	...	0.65	0.68	1.00																			
150 06	...	...	0.74	0.73	0.87	1.00																		
150 07	...	...	0.69	0.69	0.84	0.85	1.00																	
150 08	...	...	0.65	0.75	0.78	0.85	0.90	1.00																
150 09	...	...	0.60	0.66	0.73	0.73	0.83	0.87	1.00															
150 10	...	...	0.59	0.62	0.57	0.67	0.74	0.80	0.83	1.00														
150 11	...	...	0.45	0.45	0.42	0.56	0.55	0.64	0.66	0.75	1.00													
150 12																								
150 13																								
150 14																								
150 15																								
150 16	...	...	0.77	0.72	0.68	0.74	0.76	0.76	0.74	0.78	0.70	...	...	...	1.00									
150 17	...	...	0.66	0.65	0.71	0.69	0.70	0.72	0.73	0.75	0.58	...	...	...	0.90	1.00								
150 18	...	...	0.68	0.68	0.65	0.65	0.70	0.70	0.73	0.75	0.65	...	...	...	0.92	0.91	1.00							
150 19	...	...	0.58	0.58	0.50	0.52	0.57	0.57	0.68	0.69	0.59	...	...	...	0.81	0.81	0.92	1.00						
150 20	...	...	0.68	0.58	0.66	0.70	0.68	0.63	0.63	0.65	0.55	...	...	...	0.88	0.81	0.86	0.81	1.00					
150 21	...	...	0.43	0.43	0.46	0.48	0.57	0.59	0.63	0.55	0.78	...	...	...	0.57	0.50	0.57	0.51	0.53	1.00				
150 22	...	...	0.55	0.48	0.95	---	0.96	0.96	0.90	---	---	...	...	...	0.75	---	---	---	---	---	1.00			
150 23																								

Glacier Bay Oceanographic Monitoring Program Analysis of Observations, 1993–2009
Appendix G. Tables of temperature and salinity correlations between stations at select depth levels.

DEPTH	STN	00	01	02	03	04	05	06	07	08	09	10	11	12	13	14	15	16	17	18	19	20	21	22	23
200	00	...																							
200	01	...	...																						
200	02	...	...	...																					
200	03	...	...	...	...																				
200	04	...	...	...	...	1.00																			
200	05	...	...	...	...	0.87	1.00																		
200	06	...	...	...	...	0.81	0.82	1.00																	
200	07	...	...	...	...	0.83	0.90	0.90	1.00																
200	08	...	...	...	...	0.83	0.89	0.93	0.95	1.00															
200	09	...	...	...	...	0.77	0.87	0.90	0.92	0.96	1.00														
200	10	...	...	...	...	0.80	0.81	0.87	0.90	0.90	0.90	1.00													
200	11	...	...	...	...	0.76	0.79	0.73	0.83	0.84	0.85	0.88	1.00												
200	12	...	...	...	...	0.58	0.63	0.58	0.71	0.70	0.74	0.76	0.81	1.00											
200	13	...	...	...	...	...	...	...	...	...	...	...	...	...	...										
200	14	...	...	...	...	...	...	...	...	...	...	...	...	...	...	...									
200	15	...	...	...	...	0.83	0.81	0.78	0.83	0.82	0.82	0.87	0.90	0.77	...	...	1.00								
200	16	...	...	...	...	0.00	0.00	0.00	0.00	-0.00	0.00	0.00	0.00	0.00	...	...	0.00	0.00							
200	17	...	...	...	...	0.68	0.73	0.71	0.68	0.64	0.67	0.76	0.65	0.44	...	...	0.84	...	1.00						
200	18	...	...	...	...	0.74	0.71	0.68	0.69	0.70	0.69	0.84	0.79	0.65	...	...	0.88	...	0.82	1.00					
200	19	...	...	...	...	...	...	...	...	...	...	...	...	...	...	...	...	...	...	...	...				
200	20	...	...	...	...	...	...	...	...	...	...	...	...	...	...	...	...	...	...	...	...	1.00			
200	21	...	...	...	...	...	...	...	...	...	...	...	...	...	...	...	...	...	...	...	...	...	...		
200	22	...	...	...	...	...	...	...	...	...	...	...	...	...	...	...	...	...	...	...	...	...	...	...	
200	23	...	...	...	...	...	...	...	...	...	...	...	...	...	...	...	...	...	...	...	...	...	...	...	...

DEPTH	STN	00	01	02	03	04	05	06	07	08	09	10	11	12	13	14	15	16	17	18	19	20	21	22	23
250	00	...																							
250	01	...	...																						
250	02	...	...	...																					
250	03	...	...	...	...																				
250	04	...	...	...	...	1.00																			
250	05	...	...	...	...	0.86	1.00																		
250	06	...	...	...	...	0.86	0.90	1.00																	
250	07	...	...	...	...	0.87	0.92	0.93	1.00																
250	08	...	...	...	...	0.77	0.92	0.88	0.93	1.00															
250	09	...	...	...	...	0.83	0.90	0.90	0.94	0.95	1.00														
250	10	...	...	...	...	0.83	0.88	0.89	0.89	0.84	0.91	1.00													
250	11	...	...	...	...	0.83	0.89	0.86	0.90	0.83	0.92	0.96	1.00												
250	12	...	...	...	...	0.66	0.67	0.61	0.71	0.60	0.73	0.82	0.86	1.00											
250	13	...	...	...	...	...	...	...	...	...	...	...	...	...	...										
250	14	...	...	...	...	...	...	...	...	...	...	...	...	...	...	...									
250	15	...	...	...	...	...	...	...	...	...	...	...	...	...	...	...	...								
250	16	...	...	...	...	0.83	0.83	0.79	0.85	0.80	0.82	0.88	0.87	0.71	...	...	...	1.00							
250	17	...	...	...	...	...	...	...	...	...	...	...	...	...	...	...	...	...	...						
250	18	...	...	...	...	...	...	...	...	...	...	...	...	...	...	...	...	...	...	...					
250	19	...	...	...	...	...	...	...	...	...	...	...	...	...	...	...	...	...	...	...	...				
250	20	...	...	...	...	...	...	...	...	...	...	...	...	...	...	...	...	...	...	...	...	...			
250	21	...	...	...	...	...	...	...	...	...	...	...	...	...	...	...	...	...	...	...	...	...	...		
250	22	...	...	...	...	...	...	...	...	...	...	...	...	...	...	...	...	...	...	...	...	...	...	...	
250	23	...	...	...	...	...	...	...	...	...	...	...	...	...	...	...	...	...	...	...	...	...	...	...	...

The Department of the Interior protects and manages the nation's natural resources and cultural heritage; provides scientific and other information about those resources; and honors its special responsibilities to American Indians, Alaska Natives, and affiliated Island Communities.

NPS 132/112453, January 2012

www.ingramcontent.com/pod-product-compliance
Lightning Source LLC
Chambersburg PA
CBHW080245180526
45167CB00006B/2428